Recent Advances in Global Meliponiculture

Shamsul Bahri Abd Razak
Universiti Malaysia Terengganu, Malaysia

Tuan Zainazor Tuan Chilek
Universiti Malaysia Terengganu, Malaysia

Jumadil Saputra
Universiti Malaysia Terengganu, Malaysia

A volume in the Advances in
Environmental Engineering and
Green Technologies (AEEGT) Book
Series

Published in the United States of America by
IGI Global
Engineering Science Reference (an imprint of IGI Global)
701 E. Chocolate Avenue
Hershey PA, USA 17033
Tel: 717-533-8845
Fax: 717-533-8661
E-mail: cust@igi-global.com
Web site: http://www.igi-global.com

Library of Congress Cataloging-in-Publication Data

Names: Shamsul Bahri Abd Razak, editor. | Tuan Zainazor Tuan Chilek editor. | Saputra, Jumadil, editor.
Title: Recent advances in global meliponiculture / edited by Shamsul Bahri Abd Razak, Tuan Zainazor Tuan Chilek, Jumadil Saputra.
Description: Hershey, PA : Engineering Science Reference, [2023] | Includes bibliographical references and index. | Summary: "To disseminate information on stingless bee To share recent works on stingless bee To foster global network on stingless bee research"-- Provided by publisher.
Identifiers: LCCN 2022045885 (print) | LCCN 2022045886 (ebook) | ISBN 9781668462652 (hardcover) | ISBN 9781668462669 (paperback) | ISBN 9781668462676 (ebook)
Subjects: LCSH: Bee culture. | Bees.
Classification: LCC SF523 .R43 2023 (print) | LCC SF523 (ebook) | DDC 638/.1--dc23/eng/20220928
LC record available at https://lccn.loc.gov/2022045885
LC ebook record available at https://lccn.loc.gov/2022045886

This book is published in the IGI Global book series Advances in Environmental Engineering and Green Technologies (AEEGT) (ISSN: 2326-9162; eISSN: 2326-9170)

British Cataloguing in Publication Data
A Cataloguing in Publication record for this book is available from the British Library.

All work contributed to this book is new, previously-unpublished material.
The views expressed in this book are those of the authors, but not necessarily of the publisher.

For electronic access to this publication, please contact: eresources@igi-global.com.

Advances in Environmental Engineering and Green Technologies (AEEGT) Book Series

ISSN:2326-9162
EISSN:2326-9170

Editor-in-Chief: Sang-Bing Tsai, Zhongshan Institute, University of Electronic Science and Technology of China, China & Wuyi University, China; Ming-Lang Tseng, Lunghwa University of Science and Technology, Taiwan; Yuchi Wang, University of Electronic Science and Technology of China Zhongshan Institute, China

MISSION

Growing awareness and an increased focus on environmental issues such as climate change, energy use, and loss of non-renewable resources have brought about a greater need for research that provides potential solutions to these problems. Research in environmental science and engineering continues to play a vital role in uncovering new opportunities for a "green" future.

The **Advances in Environmental Engineering and Green Technologies (AEEGT)** book series is a mouthpiece for research in all aspects of environmental science, earth science, and green initiatives. This series supports the ongoing research in this field through publishing books that discuss topics within environmental engineering or that deal with the interdisciplinary field of green technologies.

COVERAGE

- Waste Management
- Sustainable Communities
- Air Quality
- Alternative Power Sources
- Green Transportation
- Renewable Energy
- Cleantech
- Pollution Management
- Water Supply and Treatment
- Industrial Waste Management and Minimization

IGI Global is currently accepting manuscripts for publication within this series. To submit a proposal for a volume in this series, please contact our Acquisition Editors at Acquisitions@igi-global.com or visit: http://www.igi-global.com/publish/.

Titles in this Series

For a list of additional titles in this series, please visit:
https://www.igi-global.com/book-series/advances-environmental-engineering-green-technolo-
gies/73679

Implications of Nanoecotoxicology on Environmental Sustainability
Rafiq Lone (Central University of Kashmir, Ganderbal, Jammu and Kashmir, India) and
Javid Ahmad Malik (Guru Ghasidas Vishwavidyalaya, Bilaspur Chhattisgarh, India)
Engineering Science Reference • copyright 2023 • 300pp • H/C (ISBN: 9781668455333)
• US $250.00 (our price)

Food Sustainability, Environmental Awareness, and Adaptation and Mitigation Strategies
for Developing Countries
Ahmad Ni'matullah Al-Baarri (Diponegoro University, Indonesia) and Diana Nur Afifah
(Diponegoro University, Indonesia)
Engineering Science Reference • copyright 2023 • 315pp • H/C (ISBN: 9781668456293)
• US $110.00 (our price)

Biomass and Bioenergy Solutions for Climate Change Mitigation and Sustainability
Ashok Kumar Rathoure (M/s Akone Services, India) and Shankar Mukundrao Khade
(Ajeenkya D.Y. Patil University, India)
Engineering Science Reference • copyright 2023 • 412pp • H/C (ISBN: 9781668452691)
• US $250.00 (our price)

Transcending Humanitarian Engineering Strategies for Sustainable Futures
Yiannis Koumpouros (University of West Attica, Greece) Angelos Georgoulas (University
of West Attica, Greece) and Georgia Kremmyda (University of Warwick, UK)
Engineering Science Reference • copyright 2023 • 315pp • H/C (ISBN: 9781668456194)
• US $270.00 (our price)

Handbook of Research on Building Greener Economics and Adopting Digital Tools in the
Era of Climate Change
Patricia Ordóñez de Pablos (The University of Oviedo, Spain)
Engineering Science Reference • copyright 2022 • 391pp • H/C (ISBN: 9781668446102)
• US $350.00 (our price)

For an entire list of titles in this series, please visit:
https://www.igi-global.com/book-series/advances-environmental-engineering-green-technolo-
gies/73679

701 East Chocolate Avenue, Hershey, PA 17033, USA
Tel: 717-533-8845 x100 • Fax: 717-533-8661
E-Mail: cust@igi-global.com • www.igi-global.com

Table of Contents

Detailed Table of Contents

Nik Nadia Syamimi Mat, Universiti Malaysia Terengganu, Malaysia
Norizah Mhd Sarbon, Universiti Malaysia Terengganu, Malaysia

Marinade is a mixture of non-meat ingredients in the form of a liquid solution applied to raw meat to delay the activity of bacteria and enzymes. Traditional marinade commonly uses salt as an ingredient. However, consuming a high intake of salt may lead to health problems. Traditional marinade only focused on prolonging the shelf life of the meat. Hence, this review aims to provide an overview of the recent advances on the application of honey in marinades towards hetero-cyclic amines formation and physicochemical and sensory properties of marinated products in detail. The physicochemical and sensory properties of various marinades have been thoroughly discussed. The results indicated that honey marinade showed better properties compared to other ingredients such as sugar, salts, and lactic acid in terms of formation of HCA, chemical, physical, and sensory properties. This chapter offers an overview of the recent advances in the application of honey in marinades in the meat industry.

Stingless bees (Hymenoptera, Apidae, Meliponini) are common pollinators in the Malaysian agricultural ecosystem. Stingless bees are regarded as a good candidate for commercial pollination because of their specialized foraging adaptations and frequent visitation to cultivated fields. Unlike honeybees and bumble bees, stingless bees have not yet been commercially bred on a large scale for pollination purposes. Several studies outside Malaysia have shown that stingless bees' foraging activities may increase the production and quality of fruits. However, the role of stingless bees in producing quality fruits in open fields or in greenhouse crops in the Malaysian agricultural ecosystem is still unknown. In this review, the authors discuss the efficiency of stingless bees, Heterotrigona itama, pollination services on some important cultivated crops in Malaysia, namely chili (Capsicum annuum), cucumbers (Cucumis sativus), and rock melon (Cucumis melo) based on previous reports. The findings revealed that pollination by H. itama can increase fruit size and weight, seed number, and pericarp volume.

Chapter 3

In Thailand, there have been limited investigations on the antibacterial properties of stingless bee honey. The purpose of this research is to investigate the physicochemical and antibacterial characteristics of five stingless bee species, including Lepidotrigona flavibasis, L. doipaensis, Lisotrigona furva, Tetragonula laeviceps species complex, and T. testaceitarsis complex from two geographical locations in Thailand: North (Chiang Mai) and Southeast (Chanthaburi). The moisture content from five species of stingless bee ranged from 27.6 to 32.0 g/100g. The range of pH in stingless bee honey was 3.5 to 3.8, which is slightly lower than the pH of Apis mellifera honey.

The total acidity of stingless bee honey ranged from 44.0 to 216.9 meq/kg. The antimicrobial property of honey samples was investigated by the agar disc-diffusion method followed by MIC/MBC assay. Notably, with the exception of L. furva, stingless bee honeys were shown to exhibit antibacterial against the Gram-negative bacteria greater than Gram-positive bacteria.

Chapter 4
    Ali Agus, Universitas Gadjah Mada, Indonesia
    Agussalim Agussalim, Faculty of Animal Science, Universitas Gadjah
        Mada, Indonesia

Domestication and propagation in stingless bees is called meliponiculture. The aims of meliponiculture are to make it easier to control the colonies health and development and to make it easy when harvesting stingless bee products (honey, bee bread, and propolis), furthermore, for advanced study and development like multiple colonies, to produce honey, bee bread, and propolis. Therefore, this paper focuswa on stingless bees Tetragonula laeviceps: the domestication and propagation technique, production of stingless bee products (honey, bee bread, and propolis), the daily activity of workers (foragers), the chemical composition (glucose, fructose, sucrose, reducing sugar, moisture, protein, ash, phenolic, flavonoid, vitamin C, antioxidant activity, minerals content, and amino acids) of honey from T. laeviceps, the pests and the challenges in meliponiculture of stingless bees.

Chapter 5
    Leo Grajo, Grajo's Farm, Philippines

The Bicol Region is the birthplace of meliponiculture in the Philippines using the native stingless bee species, Tetragonula biroi Friese. Mr. Rodolfo Palconitin of Guinobatan, Albay, started the traditional method of stingless beekeeping using indigenous material, the coconut shell, which he called bao tech or coconut shell technology. It is a form of natural hive duplication wherein coconut shell halves are gradually mounted on top of each other as the colony grows. In this technology, hive product harvesting and colony splitting are done when the stingless bees have filled up the coconut shell halves. Inspired by the visit to the University of Los Baños (UPLB) Bee Program in 2010, the Grajo's Farm started using bao technology with several experimental hives upon return to home.

 Fisal Haji Ahmad, Universiti Malaysia Terengganu, Malaysia
 Mohd Amiruddin Abdul Wahab, Universiti Malaysia Terengganu, Malaysia
 Tuan Zainazor Tuan Chilek, Universiti Malaysia Terengganu, Malaysia
 Amir Izzwan Zamri, Universiti Malaysia Terengganu, Malaysia
 Shamsul Bahri Abd Razak, Universiti Malaysia Terengganu, Malaysia
 Azril Dino Abd Malik, Naluri Pantas Sdn. Bhd, Malaysia

Generally, there are two types of beekeeping: the Apini tribe and the Meliponini tribe. Both tribes produce honey and have a good demand due to their health benefit properties. Considering the influence of diverse factors on honey composition and the lack of studies, establishing quality standards for stingless bee honey (Meliponini tribe) is still challenging and need to do to protect the consumer. In this sense, this study aimed to determine the total soluble protein content and compare the SDS-PAGE profile between two species of Apini tribe and two species of Meliponini tribe. Protein concentrations in honey samples were varied and resulted in a micro component in honey. SDS-PAGE profile for Meliponini tribe showed more number of protein bands compared to protein from Apini tribe. The unique protein bands that appeared in the Meliponini tribe may have potential as a biomarker to justify the authenticity and quality of that honey, which is known as Unique Kelulut Factor (UKF).

 Mannur Ismail Shaik, Universiti Malaysia Terengganu, Malaysia
 Noor Zulaika Zulkifli, Universiti Malaysia Terengganu, Malaysia
 Jaheera Anwar Sayyed, Universiti Malaysia Terengganu, Malaysia
 John Sushma Nannepaga, Sri Padmavati Mahila Visvavidyalayam, India
 Guruswami Gurusubramanian, Mizoram University, India
 Shamsul Bahri Abd Razak, Universiti Malaysia Terengganu, Malaysia

Honey is a natural product produced from the nectar of a variety of plants by stingless bees. Honey has been utilized for nutritional food for ages, and in recent years, stingless bee honey has been exploited as a food supplement for excellent health, cosmetic maintenance, and culinary enjoyment. Stingless bee honey has a higher moisture content (30-40%) and acidity than honeybee honey due to the presence of organic acid, mineral, and other trace components. Honey's moisture content is a key aspect that affects its stability and shelf life. The current study aimed to accesses the quality of dehydrated stingless bee honey from three different species namely,

Heterotrigona itama, Geniotrigona thoracica, and Tetrigona apicalis. The dehydration treatment of T1 (20% moisture content at 60oC in 6 hours) and T2 (15% moisture content at 60oC in 8 hours) honey samples were subjected to physicochemical properties and microbial population studies.

Chapter 8

Amir Izzwan Zamri, Universiti Malaysia Terengganu, Malaysia
Nor Hazwani Mohd Hasali, Universiti Malaysia Terengganu, Malaysia
Muhammad Hariz Mohd Hasali, Universiti Malaysia Terengganu, Malaysia
Tuan Zainazor Tuan Chilek, Universiti Malaysia Terengganu, Malaysia
Fisal Ahmad, Universiti Malaysia Terengganu, Malaysia
Mohd Khairi Mohamed Zainol, Universiti Malaysia Terengganu, Malaysia

The study was to compare and evaluate the performance of stingless bee honey (Heterotrigona itama spp.) with ordinary honey in terms of proximate composition as a comparison. Both honeys have shown diverse application and importance either traditionally and scientifically. However, due to the heightened interest on stingless bee honey, antimicrobial tests were also performed to determine the inhibition activity of stingless bee honey against food-borne pathogens using agar well diffusion assay. All three honey samples showed very good inhibitory activities (measured by inhibition zone) against Salmonella typhimurium (25-33 mm), Escherichia coli (17-33 mm), Pseudomonas aeruginosa (15-25 mm), and Staphylococcus aureus (25-29 mm). As for resistance to bile salts, pH tolerance was done and indicated the Lactic acid bacteria was able to survive the human digestive system. The haemolytic study shows that the LAB used was not virulent when introduced to red blood cells, which is important for any bacterium to be classified as safe.

Chapter 9

Suhaila Ab Hamid, Universiti Sains Malaysia, Malaysia

Insects occur in large numbers. Therefore, it is important to have a system to identify the different species of insects. Traditional morphological identification of insects requires an experienced entomologist while molecular techniques require laboratory expertise and involve substantial costs. Due to that, there has been a dramatic increase of studies using morphometric analysis in understanding the systematics, taxonomy, and diversity of stingless bees. Morphometric analysis is a powerful tool as it is effective with minimum technical experience. It is a simple technique because of the current availability of cheap computer technology equipped with software, and at the same time, this method preserves the physical integrity of the shape measured. Morphometric analysis makes it credible to recognise morphological disparity and lead ways to explore the causes, both within and between, the stingless bee populations.

Meliponiculture is the practice of handling the stingless bee for a lot of beneficial products including honey, bee bread, and propolis. Heterotrigona itama is one of the most cultured species by bee keepers in Malaysia. This research objective was pollen identification from pollen pot of H. itama. Five colonies of H. itama were observed from September 2014 until August 2015 for the foraging and pollen collection. Meanwhile, for palynology studies, the pollen sampling was done for four periods, which are September 2014, December 2014, March 2015, and June 2015. from three districts: Jeli, Kota Bharu, and Tanah Merah. One Way ANOVA was conducted, and results showed significant difference, $p<0.05$ for pollen area, pollen height of pot, pollen diameter, honey area, honey, honey diameter pot, honey height of pot, honey number of pot, and amount of honey. A total of 17,097 pollen were counted based on 66 species of identified pollen within 37 families. There was significant difference between locations and sampling period. The different geographical ranges determine various types of pollen.

The various botanical origins may be influenced by the type of plant used as a food source, which affects the chemical composition of propolis. The purpose of this work was to determine the antioxidant activity, total phenolic content (TPC), and total flavonoid content (TFC) of propolis extracted from Indo-Malayan stingless bees, Heterotrigona itama, rearing at different botanical regions. Propolis was obtained from two different botanical origins: Forested area (H. Itama-FA) propolis from Taman Pertanian Sekayu, Terengganu and Hevea brasiliensis (HB) propolis from stingless bees that reared in the rubber smallholding at Bukit Berangan, Terengganu (H. Itama-HB). TPC and TFC concentrations were evaluated using a UV-Vis Spectrophotometer, whereas antioxidant activity was determined using the DPPH free radical assay method. The results showed that the propolis of stingless bees rearing in rubber smallholdings area and the wildly available in forest area have comparable quality in terms of promising sources of antioxidant compounds.

Chapter 12

Shamsul Bahri Abd Razak, Universiti Malaysia Terengganu, Malaysia
Muhammad Izzhan, Universiti Malaysia Terengganu, Malaysia
Nur Aida Hashim, Universiti Malaysia Terengganu, Malaysia
Norasmah Basari, Universiti Malaysia Terengganu, Malaysia

Stingless bee farming or meliponiculture is a flourishing industry in Malaysia. The common practice by local stingless bee keepers in order to get new colonies is to obtain feral stingless bees hive from their natural habitat. This practice includes cutting down whole trees to extract stingless bee colonies for domestication. This is not a sustainable way of meliponiculture. The more efficient, sustainable, economic, and eco-friendly method is to breed stingless bees in hives and propagate them for colony multiplication. The aim of this experiment is to provide a good propagation method for stingless bee (Heterotrigonaitama) by dividing brood and queen cells and transfer them into a new box (split method). This method requires a portion of brood with queen cells from original log hive to be transferredinto an empty box hive. From this experiment, 80% of new box hives become new colonies (with new queens).

Chapter 13

Siti Salmah, Universitas Andalas, Indonesia
Henny Herwina, Universitas Andalas, Indonesia
Jasmi Jasmi, College of Health Sciences Indonesia, Indonesia
Idrus Abbas, Universitas Andalas, Indonesia
Dahelmi Dahelmi, Universitas Andalas, Indonesia
Muhammad N. Janra, Universitas Andalas, Indonesia
Buti Yohenda Christy, Universitas Andalas, Indonesia

This chapter summarizes the works on Sumatran bees from three research periods: between 1980-1987 on several locations in West Sumatra, 1990 at Kerinci Seblat National Park, and between 2019-2020 at some beekeepers in West Sumatra. In total, there were 27 stingless bee species, one stingless bee forma (Tetragonula minangkabau forma darek), and three honey bee species identified. Most of these stingless bee and honey bee species inhabit the Sumatran lowland primary forest. There were four patterns of species distribution observed in this study: rare species that were confined to primary forest, moderate or abundant species that were bound to primary forest, species that inhabited both primary and secondary forest, and species that adapt to disturbed areas. Apis andreniformis, A. dorsata, A. cerana indica, Heterotrigona itama, Sundatrigona moorei, Tetragonula fuscobaltealta, T. drescheri, T. laeviceps, and T. minangkabau were example of adaptive species.

Preface

Stingless bee is a large group of bees from the *Apidae* family and order *Hymenoptera*. Approximately 700 species of stingless bees were recorded, most found in tropical countries. Each species has unique characteristics regarding morphology and behaviour, including size, population and habitat quality. Stingless bees are highly eusocial insects. They possessed stingers but were highly reduced, which rendered them unusable for defence. Therefore, they are safe for domestication as they do not sting.

In addition, most species are perennial, which suited well for meliponiculture. The drone of a stingless bee can be seen from a distinctive congregation of up to several hundred individuals, which can persist several times. Generally, male production in a social insect colony is influenced by outside factors related to climatic periodicity and factors inside the colony, such as colony strength and demographic composition.

In varying situations, it is essential to enhance understanding about the global meliponiculture, especially their economic contribution toward sustaining the ecology and environment, pollination services by stingless bees, their honey unique properties for consumer's wellbeing, domestication and propagation of stingless bee, hive design and innovation, physicochemical properties and microbial population of stingless bee, microbiological diversity and properties, morphometric analysis, palynology, phenolic and flavonoids content of propolis extracts, propagation of stingless bee.

Awareness and public interest to domesticate stingless bee has been steadily growing in many native stingless bee countries. Countries such as Malaysia, Thailand, Philippine, and Indonesia of which stingless bee can be found in abundance are actively developing protocols for sustainable domestication of stingless bee. Some focus areas in meliponiculture such as innovation in hive design has been a priority activity by researchers and stingless beekeepers.

Honey quality is one of the major concerns by consumers. Physicochemical analysis of honey in accordance with standards from various authorities from each country is carried out in ensuring that honey produced by stingless bee are of highest quality. Various aspects of meliponiculture which are presented in this book will satisfy some of the major concerns of stingless beekeprs, researchers, students and consumers. This is in line with the increasing demand for their honey.

INSIDE THIS BOOK

In this regard, the book is pathbreaking by reviewing of honey application in physicochemical and sensory properties of marinated products. Traditional marinade commonly uses salt as an ingredient but consuming a high intake of salt may lead to health problems. The physicochemical and sensory properties of various marinades have been thoroughly discussed in this chapter.

In the second chapter, A review on the pollination services by stingless bees, *Heterotrigona itama* (*Hymenoptera, Apidae, Meliponini*) on some important crops in Malaysia is reported. Stingless bees are a good candidate for commercial pollination because of their specialized foraging adaptations and frequent visitation to cultivated fields. Unlike honeybees and bumble bees, stingless bees have not yet been commercially bred on a large scale for pollination purposes.

The third chapter provides a deeper understanding of the antimicrobial activity from five species of stingless bee (*Apidae meliponini*) honey from South East Asia. There have been limited investigations on the antibacterial properties of stingless bee honey. This research investigates the physicochemical and antibacterial characteristics of five stingless bee species, including *Lepidotrigona flavibasis, L. doipaensis, Lisotrigona furva, Tetragonula laeviceps* species complex and *T. testaceitarsis* complex, from two geographical locations in Thailand.

The fourth chapter stresses the approach for the domestication and propagation of stingless bee. The aims of meliponiculture are to make it easier to control the colonies health and development, easy in harvesting stingless bee products (honey, bee bread, and propolis). Advanced study and development such as multiple colonies, to produce honey, bee bread, and propolis is also presented.

The fifth chapter refers to Brazil-Inspired vertical hive technology for the Philippine version. Bicol Region is the birthplace of meliponiculture in the Philippines using the native stingless bee species. Mr Rodolfo Palconitin of Guinobatan, Albay, started the traditional method of stingless beekeeping using indigenous material – the coconut shell, which he called Bao Tech or Coconut shell technology. Inspired by the visit to the University of Los Baños (UPLB) Bee Program in 2010, Grajo's Farm started using Bao Technology with several experimental hives upon return to his home.

The sixth chapter compared the total soluble protein content and SDS-PAGE pattern between four different types of honey. Two types of beekeeping: the Apini tribe and Meliponini tribe. Both tribes produce honey and have a good demand due to their health benefits. Considering the influence of diverse factors on honey composition and the lack of studies, establishing quality standards for stingless bee honey (Meliponini tribe) is still challenging but necessary to protect the consumer's interest. In this sense, protein concentrations in honey samples were varied and resulted in a micro component in honey.

The seventh chapter examined the dehydration treatment effect on physicochemical properties and microbial population of stingless bee honey from three different species. Honey has been utilized for nutritional food for ages, and in recent years, stingless bee honey has been exploited as a food supplement for excellent health, cosmetic maintenance, and culinary enjoyment. Stingless bee honey has a higher moisture content (30-40%) and high acidity than honeybee honey due to the presence of organic acid, minerals, and other trace components. The moisture content of honey is an important factor contributing to its stability and shelf-life. Honey's moisture content is a key aspect that affects its stability and shelf life.

The eighth chapter compared the performance of stingless bee honey (*Heterotrigona itama spp.*) with apis honey in terms of proximate composition as a comparison. Both honeys have shown diverse applications and importance, either traditionally or scientifically. Due to the high interest in stingless bee honey, antimicrobial tests were also performed to determine the inhibition activity of bee honey against food-borne pathogens using an agar well diffusion assay.

The ninth chapter identify the morphometric analysis in stingless bees (*Apidae meliponini*) diversity. It important to have a system for identifying the different species of insects. Traditional morphological identification of insects requires an experienced entomologist, while molecular techniques require laboratory expertise and involve substantial costs. Due to that, there has been a dramatic increase in studies using morphometric analysis to understand stingless bees' systematics, taxonomy, and diversity.

The tenth chapter examined the pollen identification from the pollen pot of *H. itama*. The study found a significant difference for pollen area, pollen height of pot, pollen diameter, honey area, honey diameter pot, the honey height of the pot, honey number of pot and the amount of honey.

The eleventh chapter determined the antioxidant activity, total phenolic content (TPC), and total flavonoid content (TFC) of propolis extracted from Indo-Malaya stingless bees: *Heterotrigona itama*, rearing at different botanical regions. Propolis was obtained from two different botanical origins, Forested area (H.Itama-FA) propolis from Taman Pertanian Sekayu, Terengganu and *Hevea brasiliensis* (HB) propolis from stingless bees that reared in the rubber smallholding at Bukit Berangan, Terengganu (H.Itama-HB).

The twelfth chapter explored the propagation of stingless bee using a colony split-technique for sustainable meliponiculture. The common practice by local stingless beekeepers to get new colonies is to obtain feral stingless bees hive from their natural habitat. This practice includes cutting down whole trees to extract stingless bee colonies for domestication. It is not a sustainable way of meliponiculture. The more efficient, sustainable, economical and eco-friendly method is to breed stingless bees in hives and propagate them for colonies multiplication.

The last chapter observed stingless bees and honeybees of West Sumatra, Indonesia. Four patterns of species distribution were confined to the primary forest, moderate or abundant species that were bound to primary forest, species that inhabited both primary and secondary forest, and species that adapt to disturbed areas. *Apis andreniformis, A. dorsata, A. cerana indica, Heterotrigona itama, Sundatrigona moorei, Tetragonula fuscobaltealta, T. drescheri, T. laeviceps,* and *T. minangkabau* were example of adaptive species.

CONCLUSION

Meliponiculture in the South East Asia is considered still in its infant stage but shows a significant development in a short time of period. This is due to many collaborative efforts from stingless keepers and researchers in addressing concerning issues in meliponiculture. Various important aspects of meliponiculture such as sustainable beekeeping, hive design, quality of honey, development of honey standard, community development in marginal areas are becoming major focuses. A transdisciplinary approach that covers a wide range of meliponiculture aspect requires more effort in the near future. This is to ensure meliponiculture as a sustainable activity not only as income generating venture but also contributing to conservation of ecosystem.

Shamsul Bahri Abd Razak
Universiti Malaysia Terengganu, Malaysia

Tuan Zainazor Tuan Chilek
Universiti Malaysia Terengganu, Malaysia

Jumadil Saputra
Universiti Malaysia Terengganu, Malaysia

Chapter 1

A Review of Honey Application in Marinades Towards Hetero–Cyclic Amines (HCA) Formation:
Physicochemical and Sensory Properties of Marinated Products

Nik Nadia Syamimi Mat
Universiti Malaysia Terengganu, Malaysia

Norizah Mhd Sarbon
Universiti Malaysia Terengganu, Malaysia

ABSTRACT

Marinade is a mixture of non-meat ingredients in the form of a liquid solution applied to raw meat to delay the activity of bacteria and enzymes. Traditional marinade commonly uses salt as an ingredient. However, consuming a high intake of salt may lead to health problems. Traditional marinade only focused on prolonging the shelf life of the meat. Hence, this review aims to provide an overview of the recent advances on the application of honey in marinades towards hetero-cyclic amines formation and physicochemical and sensory properties of marinated products in detail. The physicochemical and sensory properties of various marinades have been thoroughly discussed. The results indicated that honey marinade showed better properties compared to other ingredients such as sugar, salts, and lactic acid in terms of formation of HCA, chemical, physical, and sensory properties. This chapter offers an overview of the recent advances in the application of honey in marinades in the meat industry.

DOI: 10.4018/978-1-6684-6265-2.ch001

INTRODUCTION

Marinade is defined as a mixture of non-meat ingredients, in a form of liquid solution or powder which is added to uncooked food especially meat, to enhance its flavour (Yusop et al., 2011). It is one of the popular techniques used to slow down the bacterial and enzymatic activity, and provides the tenderness, textural, and structural changes with a prolonged shelf-life of the meat product (Ozogul & Balikci, 2013). In general, marinades contain of salt, phosphates, sugar, seasonings, spices, oils, and/or acids and usually applied to a grilled, roast, fried and steam meat product (Daly et al., 2013; Viegas et al., 2015). Nowadays, many marinades product has been developed and commercialized in the market such as liquid marinade, paste marinade, and powder marinade to make it easy for the consumer. Moreover, the interest in using marinade towards the meat and fish product maybe because it can reduce the formation of Hetero-cyclic Amine in the cooked product (Jinap et al., 2015).

Hetero-cyclic amines (HCA) are mutagenic and carcinogenic compounds developed in proteinaceous foods that have been roasted, grilled, steamed, or broiled at high temperatures (Kataoka et al., 2012). It is an organic compound which contains one or more aromatic rings (Pleva et al., 2020). High intake of HCA may increase the risk of variety of common cancer such as colon, breast, prostate and colorectal (Ali et al., 2019; Fu et al., 2011). However, the formation of HCA in food products can be prevented by using marinade during food preparation (Aaslyng et al., 2016; Jinap et al., 2018). Moreover, several studies have shown that marinade has also improved the physicochemical properties of meat product (Sharedeh et al., 2015).

The physicochemical properties of marinated products are important to know the characteristics of its final product (Jambrak, 2017). The chemical (moisture content, pH, and antioxidant activity) and physical properties (cooking loss, drip loss, water-holding capacity, and texture profile) are the most analysis done towards the marinated products (Arcanjo et al., 2019; Babikova et al., 2020). Several studies also comparing the physicochemical properties of unmarinated and marinated products (Manful et al., 2020; Quelhas et al., 2010). This comparison is to know the effectiveness of the marinades towards the meat product (Mohammed et al., 2017). Moreover, the addition of the marinade in meat product not only improves the physicochemical properties but also enhance the sensory acceptability of the product (Behera et al., 2020).

Sensory properties are a scientific discipline used to evoke, quantify, evaluate, and interpret responses to certain features of products as they are experienced by the senses of sight, scent, taste, touch, and hearing (Zeng et al., 2008). It can be measured by using sensory affective test and the attributes used to measures are appearance, colour, odour, taste, texture, and overall acceptability (Tomac et al.,

2

2015). Previous study has been done on the sensory properties of marinated chicken, meat, fish, and other marinated product using different ingredients in marinade (Hafez & Eissawy, 2018; Iqbal et al., 2016; Oyeyinka et al., 2019). Besides, study using different ingredients in marinade such as cinnamon, oregano, thyme and honey is widely done (Hasnol et al., 2014; Sepahpour et al., 2018; Van Haute et al., 2016). The presence of compounds in the ingredients used will benefits towards the marinated product (Kumar et al., 2015).

Honey is a natural biological substance, composed of a special mixture of components produced by honeybees from plant secretions, which is of great value to human beings, both as medicinal products and as food products (Cheung et al., 2019; Eshete & Eshete, 2019). It is mainly consisting of sugars and water. Sugars in honey contain primarily monosaccharides and oligosaccharides (Vallianou et al., 2014). Moreover, honey also contains valuable nutrients such as minerals, vitamins, enzymes, free amino acids and various phenolic (Rao et al., 2016; Sulaiman & Sarbon, 2022). Previously, study on the effects of honey addition in marinades towards marinated product has been done. A study by Tänavots et al. (2018) focused on the physicochemical properties (cooking loss and pH measurement) of mustard-honey in pork. Other researchers studied on the physicochemical and sensory properties of roasted chicken, chicken baked without skin, grilled beef satay and fried beef steak using different types of honey-containing marinades formulation (Jinap et al., 2018; Shamsudin et al., 2020).

There were some reviews have been published on the marination. Vlahova-Vangelova and Dragoev (2014) had reviewed on the effects of marination on meat safety and human health. Additionally, seafood marination has been summarized recently by Behera et al. (2020). However, there is no scientific review on the application of honey in marinades toward hetero-cyclic amines (HCA) formation, physicochemical and sensory properties of marinated products. Therefore, this review aims to provide detail information on the recent advances on honey application in marinades towards hetero-cyclic amines formation, physicochemical and sensory properties of marinated products. The effects of application of honey in marinades are summarize, and prospects on technological developments of marinade are also discussed. It is hoped that this review could provide better understanding on the effect of honey marinade towards the quality of marinated products.

HONEY

Honey is a natural sweetener and viscous substance produced by honeybees from the nectar of flowers (Rahman et al., 2014). It contains combination of sugars (mainly fructose, glucose, and maltose) and other compounds, including oligosaccharides, minerals, carbohydrates, enzymes, and phytochemicals such as flavonoids and ferulic and caffeic acids (Iftikhar et al., 2010). These compounds give the benefit both to the products and the consumer's health (Cianciosi et al., 2018). Several studies have been performed on the impact of applying honey to food products.

Types of Honey

There are varieties of honey that have been commercialized worldwide. In Europe, there are unifloral honeys, such as black locust, sweet chestnut, lime, sage, and winter sweet honey (Svečnjak et al., 2015). Besides, a tropical country such as Malaysia are rich in various types of honey such as Gelam, Acacia, Borneo, Kelulut, Pineapple, and Tualang honeys (Chong et al., 2017). However, the most common types of honey found in Malaysia are Acacia, Kelulut, and Tualang honey (Sulaiman & Sarbon, 2022).

Acacia honey is a honey collected from the nectar of Acacia mangium trees, which produced by Apis mellifera (Zolkapli et al., 2018). It is available worldwide with different physical and chemical characteristics depending on its geographical origin (Rahaman et al., 2013). Acacia honey is popular because of its permanently liquid physical condition, light colour, floral scent, sweet, and delicate fragrance (Schievano et al., 2020). Since Acacia honey is available worldwide, there are now a number of published papers detailing the physicochemical properties of Acacia honey from various countries of origin, Malaysia, Saudi Arabia, Benin and Australia (Ahamed et al., 2017; Azonwade et al., 2018; Zolkapli et al., 2018).

Kelulut honey is a honey produced by small size stingless bees, Trigona spp or known as 'Kelulut' which usually form a complex social colony (Kek et al., 2014). Kelulut honey is darker in colour, more fluids in texture, undergoes slow crystallisation, and slightly sour taste (Budin et al., 2017; Kek et al., 2014; Ngoi, 2016). There are many works focused on the physicochemical properties of Kelulut honey from different states in Malaysia such as Kelantan, Terengganu, Johor, and Sarawak (Dan et al., 2018; Fatima et al., 2018; Wong et al., 2019).

Tualang honey is a multi-floral jungle honey of Asian origin and is produced by Apis dorsata bee species (Sarfraz Ahmed & Othman, 2017; Khalil et al., 2015). It is collected from the combs of Asian rock bees (Apis dorsata), which construct their hives high up in the Tualang tree *(Koompassia excelsa)* (Tan et al., 2009). There are more than 100 nests may be found on one Tualang tree and a tree can yield about 450 kg of honey. Tualang honey is dark brown in colour and a bit sour compared to Gelam honey (Sarfarz Ahmed & Othman, 2013; Sulaiman & Sarbon, 2022). Several researchers interested in studying of the physicochemical properties, total phenolic, and flavonoid contents, antioxidants, and antifungal properties in Tualang honey (Shehu et al., 2016; Sulaiman & Sarbon, 2022).

Physicochemical Properties of Honey

Chemical Properties of Honey

Honey has several chemical properties which contribute to its composition and taste (Eteraf-Oskouei & Najafi, 2013). It is strongly linked to its botanical source, environmental conditions, bee species involved in its production geographical area, and storage conditions (Ávila et al., 2019). Chemical properties of honey include moisture content, pH, hydroxymethylfurfural (HMF) content, and total acidity (Ahamed et al., 2017; Dan et al., 2018; Sulaiman & Sarbon, 2022). Moisture content is the most important parameter for the evaluation of honey. It may vary based on air humidity during storage (Prica et al., 2014). Honey is hydroscopic and can absorb moisture from the environment (Uckun & Selli, 2017). Moisture content of three types of honey samples (Tualang, Gelam, and Acacia) from Malaysia were studied by Rahaman et al. (2013). The results indicated that Acacia honey have the lowest moisture content compared to Tualang and Gelam honey; 19.53%, 24.07%, 25.20%, respectively. Tualang and Gelam honey contained high moisture content, which exceeding 20%, while the moisture content of Acacia honey was within the permitted ranged (World Health Organization, 2001).

However, as Malaysia is a tropical country with rainy season all over the year, this moisture content (<30%) is acceptable (Dan et al., 2018). Moreover, high moisture content of honey will quickly contribute to honey fermentation due to osmotolerant yeasts (Kek et al., 2017; Sulaiman & Sarbon, 2022). The fermentation process results in the production of alcohol. Alcohol in the presence of oxygen would then break down into lactic acid and water thereby influencing the pH of honey and provide the sour taste to the honey (Prica et al., 2014; Sulaiman & Sarbon, 2022). pH is an important parameter to predict the quality of honey. pH of honey is varying between 3.2 and 4.5 (Karabagias et al., 2014; Sulaiman & Sarbon, 2022). It is important during the extraction and storage of honey because pH affects its texture, stability, and shelf life (Kek et al., 2017). Honey has slight acidity; it is due to the fact that honey contains around 0.57% organic acid (De Silva et al., 2016). Main organic acid is gluconic acid, while low amount of organic acids such as formic, acetic and citric have been found in honey (Rahman et al., 2014). Chan et al. (2017) had studied on the pH of Kelulut honey from three different region in Malaysia. The results indicated that ranged pH was from 3.29 to 3.71.

Physical Properties of Honey

Physical analysis is used to measure the quality and authenticity of honey (Dan et al., 2018). The physical properties of honey include electric conductivity, viscosity, colour intensity, total soluble solid and specific gravity (A-Rahaman et al., 2013; Ahamed et al., 2017; Villalpando-Aguilar et al., 2022). Electrical conductivity (EC) is a capability of a honey solution to conduct an electrical current. It can be determined by using a conductivity meter (A-Rahaman et al., 2013). Electrical conductivity of the honey is closely related to proteins, botanical origin, organic acid, and other components, such as polyols and sugars, which may function as electrolytes (De-Melo et al., 2018; Nordin et al., 2018).

Ya'akob et al. (2019) had compared the electrical conductivity of 10 stingless bee honey sample from various districts of Johor. The results indicated the electrical conductivity of stingless bee honey varies between 0.1 mS/cm to 0.79 mS/cm. The electrical conductivity from both studies did not exceed the limit set by Codex Alimentarius (<0.8 mS/cm) (World Health Organization, 2001). Next, viscosity is the intrinsic resistance that opposes the flow of the honey. It affects the quality of the product as well as the design of honey-processing equipment (Yanniotis et al., 2006). A study done by Yap et al. (2019) found that the viscosity of the honey raised more intensely at a higher dehydration temperature. The viscosity of the honey was 8.11, 56.21 and 279.88 Pa-s after dehydration at 40°C, 55°C and 70°C for 18 h, respectively while Chong et al. (2017) found that thermally processed honey had higher viscosities than raw honey.

The Effect of Heat Treatment on Honey Properties

Heating is one of the steps in convectional processing of honey (Pimentel-González et al., 2016). A broad range of temperatures ranging from 30°C to 140°C for a few seconds to several hours has been conducted by honey farmers worldwide (Eshete & Eshete, 2019). Heating of honey is performed at various temperature and durations from mild temperatures of 50°C to 90°C or 15 to 120 min, 50°C to 80°C for 15 to 60 min, 60°C to 100°C for 2–20 min and 75°C to 100°C for 15 to 90 min (Bodor et al., 2017; Yap et al., 2019; Zhang et al., 2012; Zhao et al., 2018). The heat treatment on honey usually done for two purposes: (a) to alter the tendency of honey to crystallize, and (b) to eliminate the microorganisms that pollute honey (Sulaiman & Sarbon, 2022). However, heating of honey will change the quality of honey. Heating damage can be shown by evaluating quality control parameters, such as HMF value and diastase activity (Sajid et al., 2019).

Hydroxymethylfurfural (HMF) is a cyclic aldehyde and it is an undesirable compound produced by sugar degradation due to the Maillard reaction (a non-enzymatic browning reaction) (Shapla et al., 2018). Formation of HMF is a natural process that is slow at room temperature, but thermal treatment of honey at high temperatures will lead to a substantial increase in HMF content (Samira, 2016). Some factors affect HMF levels are storage conditions, temperature, time of heating, pH and floral source, thereby giving an indicator of overheating and storage under poor conditions (Zarei et al., 2019). The increasing of HMF content under rising the temperature was reported by Tosi et al. (2008) with initial concentration of 5.8 increased to 32.4 mg/kg but did not reach the HMF limit of 60 mg/kg limit of HMF content set by Codex Alimentarius (World Health Organization, 2001). This result may be explained by the increase in the concentration of fructose that overcame the energy barrier and enabled the Maillard reaction to form HMF compounds (Ngoi, 2016).

Honey contains variety of enzymes, such as diastase, invertase, glucose-oxidase, phosphatase, and catalase (De-Melo et al., 2018). They are responsible for converting nectar and honeydew to honey. Among all, diastase is one of the major enzymes present in honey (Pasias et al., 2017). Diastases are a group of amylolytic enzymes that comprise α-and β-amylase. α-Amylase hydrolyses starch chains in α-δ-(1®4) linkages, forming dextrin. β-Amylase hydrolyses the starch chain at the end, contributing to maltose formation (De Silva et al., 2016). Diastase activity is the international parameter used to monitor the limit for thermal treatment of honey (Samira, 2016). Diastases are heat sensitive and can be degraded if subjected to elevate temperatures (De Silva et al., 2016). Thus, extremely low diastase activity suggests that the honey has been exposed to unfavorable high temperatures (Sak-Bosnar & Sakač, 2012). An Argentinian research group measured the diastase activity at changes in heated honeys. They found that diastase activity already decreased at the heating level of 60 °C-120 s (Bodor et al., 2017). Another study has been conducted by Czipa et al. (2019) who found that heating of honey reduced diastase activity from 25.8 DN (before heated) of initial concentration, decreased to 20.7 DN at 80 °C. The loss of diastase during heat treatment was attributed to structural differences in enzyme molecules. Heat gives kinetic energy to enzymes and contributes to permanent denaturation. Therefore, as the temperature of the heat rises, the enzymes can absorb more energy and become denatured (Tosi et al., 2008).

MARINADE IN FOOD PREPARATION

Advantage of Marination

Marinade refers to a combination of ingredients, in a dry or liquid form, which are added to uncooked foods (Inguglia et al., 2019). The functionality of marinade is depending on the ingredients used in their contents (Erge et al., 2018). It is widely known that marination gives a lot of benefits towards the meat product for example, improve the tenderness flavour and shelf life of the marinated products. Marination improves the tenderness of meat by soften the fibres of meat muscle (Ibrahim et al., 2018). Marination by adding salts can tenderize the meat by the following mechanisms: one is that high concentration of calcium ion as an important signal factor contributes to activation of calpain, which resulted in the degradation of cytoskeletal protein (Ouali et al., 2013). The other is that the direct effect of the calcium ion strength, including the repulsion between the electrical charges, leads to the mechanical fracturing of the Z-disk (Goli et al., 2014). The degradation of this protein structure such as titin was related to the loss of myofibrillar integrity, which improved meat tenderness (Li et al., 2017).

Salt is widely known to prolong the shelf-life of meat and has a beneficial impact on flavor, texture and water-binding capacity and reduces water activity. However, excessive intake of salt, will cause in health problems such as high blood pressure and an increased chance of cardiovascular disease (Kameník et al., 2017). Therefore, Lee et al. (2012) has studied the effect of partial and complete replacement of sodium chloride with potassium chloride on the texture of marinated broiler breast fillets. Five difference formulation of difference concentration NaCl/KCl/STPP has been studied. There are significant differences observed between treatments for the tenderness. The most tender meats were fillets marinated with 100% NaCl (11.1 N), whereas the toughest one was the 100% STPP (control), 12.8N values of MORSE.

Marination improves the shelf-life of the meat through the simultaneous action of salt and organic acids (Ozogul & Balikci, 2013). The reaction between the organic acids and salt increases the ionic strength and decrease the pH of marinade. Lower pH marinade has a major impact on the evolution of microbial development. Therefore, it may reduce the degradation of meat products by slows down the bacterial and enzyme activity (Lytou et al., 2018). A study about the effect of cinnamon and thyme in marinade has been done by Van Haute et al. (2016) and found out that they show potential to slow the growth of some spoilage microorganisms on meat product. Another work has been done by Birk et al. (2010) studies on the impact of organic acids marination on the survival of campylobacter jejuni (*C. jejuni*) on meat. The results indicated all the organic acids reduced *C. jejuni* very efficiently, causing a 4- to 6-log reduction in after 24 h of exposure at 4 °C.

Marinade Technology

Marinade has been used in the meat industry for several decades (O'Neill et al., 2019). Marinade method continuously changes following contemporary scientific and technological achievements. The oldest methods were immersion. It remains irreplaceable but nowadays, it has been modified with respect to satisfy market demand and enhance the nutritional content of the marinated products (Gao et al., 2015). There are several marinade technologies that have been modified from the conventional method (immersion) including tumbling and injection. Vacuum tumbling and injector needles are shown in Figure 1.

Figure 1. Vacuum tumbler and injector needles
Source: Casey et al. (2010)

Tumbling is meat spinning, falling and contacting with metal walls and paddles in a drum (Gao et al., 2015). It is used to accelerate the penetration of the marinade and allow more consistent distribution in meat. It causes a softening and rupture of the tissue structure, which results in an increase in brine sorption and protein extraction, and a subsequent increase in cooking yield (Pooona et al., 2019). In tumbling process, the meat chunks are picked up by the baffles and dropped inside the drum. Next, the gravity of muscle chunks, high friction, impact and extrusion forces are created and energy is transmitted to muscle chunks by baffles, resulting in meat deformation (Singh et al., 2019). Tumbling is a popular mechanical marinating technology for the manufacture of whole-muscle cured meat products and is typically applied to poultry meat to produce a ready-to-cook, value-added product to either a food processing plant or a supermarket or butcher shop (Alvarado & McKee, 2007; Kim et al., 2019).

A study done by Gao et al. (2015) on the effect of different tumbling marinade treatments on the moisture content and protein properties of prepared pork chops. The results indicate cooking loss (%) of marination using vacuum continuous tumbling marination, is lower than conventional static marination, 14.14%, 23.78% respectively. Another marinade technology is injection (Pérez-Juan et al., 2012). Injection process is a meat products move through an array of needles and the marinade is injected into the meat matrix (Palang, 2004). There are three techniques for injection of marinades: (1) artery pumping, in which the brine is pumped directly through the intact vascular system; (2) stitch pumping, through which the brine is injected (through a hollow needle) into various parts of the meat, in particular the thickest parts and those close to the joints; and (3) multiple injections, where the brine is injected simultaneously and automatically through a series of hollow needles which have a number of evenly spaced holes along their length (Ledward, 2003).

Injection marinade is widely applicable in meat industry as it provides more control over the process of marinating by providing an accurate quantity of brine to the beef, maintaining accuracy and consistency in the dosage of marinade products (Yusop et al., 2011). A study done by Kim et al. (2015) focused on the quality characteristics of marinated chicken breast using injection technology. They revealed that cooking yields of marinated chicken breast using injection technology higher than tumbling, 125.56% and 109.57%, respectively. The marinating and cooking yields of chicken breast could be related to the components in the brine, and the injecting process is considered as the most effective method for improving the dispersion of the brine.

Honey in Marinade

Nowadays, the application of honey in marinades in meat products has become popular all over the world. According to Batt and Liu (2012) Asian's people are often using honey in marinades. This is due to the benefits that have been carried by the honey in marinades towards the meat and meat products (Istrati et al., 2015). Final products of honey marinated products offer a good eating experience for consumers (Tänavots et al., 2018). Honey in marinades was reported to have a good impact towards the marinated products (O'Neill et al., 2019). Shin and Ustunol (2004) has conducted a study on the impact of honey-containing marinades on HCA formation in fried beef and chicken. Buckwheat and clover honeys were chosen for their high and low antioxidant capacity, respectively. For both types of honey, 3 different level of honey were added into the marinade formulation. The inhibition of total HCA formation in fried beef steaks was achieved with marinades containing buckwheat honey alone (4.5, 9.0, or 13.5 g) by 16%, 36%, and 41%, respectively while marinades containing clover honey alone (4.5, 9.0, and 13.5 g) reduced the total HCA formation by 10%, 31%, and 37%, respectively. However, there was no significant difference in the percent inhibition achieved by these concentrations of buckwheat or clover honeys.

Moreover, the increasing demand for convenience products and longer shelf-life has increased interest in the use of honey in marinades to improve food safety (Karam et al., 2019). Food safety in microbiological terms can be described as the absence of pathogenic organisms and toxins of microbiological origin in a specified quantity of food (Augustyńska-Prejsnar et al., 2019). Yücel et al. (2005) has studied the effect of honey treatment on some quality characteristics of broiler breast meat. Honey is added at concentrations of 0%, 20% and 30%. The results shown total bacteria count in breast meat marinated with honey was decreased with the increasing of percentage honey added in the marinades. Decrease in total bacteria counts was most pronounced in the 30% honey treatment (5.71 CFU/g) compared to 0% (6.29 CFU/g) and 20% of honey (6.05 CFU/g). This finding may be described as honey marinades delays the growth of bacteria and extends the shelf-life of breast meat. In conclusion, honey marinade can be used to minimise adverse effects on the consistency of the meat and to extend the shelf-life of the marinated products.

Another study has been done by Li and Gänzle (2016) has found that marination did not influence the survival of L. monocytogenes during pressure treatment but improved survival during storage. They studied the effect of hydrostatic pressure and antimicrobials marinades on survival of Listeria monocytogenes and enterohaemorrhagic Escherichia coli in beef. Formulation of honey garlic (Sugar, salt, fructose, honey powder, granulated garlic, sodium phosphates, soy sauce powder, garlic powder, caramel, calcium silicate, spices, monounsaturated vegetable oil, artificial flavour) and teriyaki marinade (Sugar, salt, soy sauce powder, sodium phosphates, flavour, caramel, garlic powder, onion powder, spices, xanthan gum, monounsaturated vegetable oil, sulphites) has been used.

HETERO-CYCLIC AMINE (HCA)

Hetero-cyclic amine (HCA) is a carcinogenic compound produce during heating process of protein rich foods (Oz & Cakmak, 2016). It was first discovered by Sugimura et al. (1997) and since then, there are more than 25 distinct HCAs have been isolated and characterized in cooked foods (Zhang et al., 2020). All HCAs have at least one aromatic and one heterocyclic structure that give them a different name of heterocyclic amines (Cheng et al., 2006). Based on their chemical structure, HCAs can be divided into two classes: amino-carbolines (ACs) and aminoimidazole-azaarenes (AIAs) (Quelhas et al., 2010). Amino-carbolines is a non-polar HCA (Chu et al., 2018). The non- polar HCA includes five-membered heterocyclic amines which contain a ring of five atoms (Sahar et al., 2016). Non-polar HCAs can be further classified into α-amino-carbolines (e.g. AαC and MeAαC), β-amino-carbolines Hetero-cyclic amines (HCAs) are naturally produced to some degree when proteinaceous food is heat-treated (Chen et al., 2020). (e.g. norharman, harman), γ-amino-carbolines (e.g. Trp-P-1, Trp-P-2), and δ-amino-carbolines (e.g. Glu-P-1, Glu-P-2). Unlike other HCAs, the β-amino-carbolines (e.g. norharman, harman) are none-mutagenic in the Ames Salmonella mutagenicity test due to lack the exocyclic amino group (Cheng et al., 2007). The formation of non-polar HCAs takes place through pyrolysis of amino acids and proteins at higher temperatures more than 300 (Alaejos & Afonso, 2011; Woziwodzka et al., 2013). Next, another HCA type is Aminoimidazole-azaarenes (AIAs).

Aminoimidazole-azaarenes (AIAs) is a polar HCA (Chu et al., 2018). The arrangement of polar amines has 2-amino-imidazo moiety and a methyl group bound to one of the nitrogens in the imidazo ring. Polar HCAs can be classified into three

11

subgroups based on the groups linked to the 2-amino-imidazo (Cheng et al., 2006). These subgroups are quinolines, quinoxalins and pyridines. The Aminoimidazole-azaarenes (AIAs) HCA are formed because of a complex reaction between creatine/creatinine, free amino acids, and sugars by Maillard at temperatures below than 300°C (Jinap et al., 2015). Harman and norharman do not have a free amino group, which does not induce mutagenicity in the Ames test (Pfau & Skog, 2004). Harman and norharman occurred in cooked meat, poultry, fish, and meat extract (Hasnol et al., 2014). The same reaction is postulated for both β-carbolines (harman and norharman) (Gibis, 2016). Besides, the formation mechanisms of other Amino-carbolines such as Trp-P 1 Trp-P 2 AaC and MeAaC are remain unknown (Szterk, 2015).

Main structure of IQ compounds is form due to the interaction between reducing sugar such as hexoses and amino acids via Strecker Degradation in the Maillard reaction to formaldehyde and pyridine or pyrazine. At the same time, creatinine is formed from creatine during the heating phase (Chen et al., 2020; Scalone et al., 2019). Then aldehyde, pyridine or pyrazine and creatinine react together to produce corresponding IQ compounds by aldol reactions. This reaction mechanism was further verified for the development of imidazoquinoxalin (IQx) and 4,8-DiMeIQx using the glucose labelling isotope (Zamora & Hidalgo, 2020). PhIP mechanism begins with the reaction of two precursors, which are phenylalanine and creatinine to form phenylacetaldehyde (Puangsombat, 2010). The phenylacetaldehyde formed interacts in an aldol reaction with creatinine to form an intermediate product. In the following condensation reaction, PhIP is produced from this substance as the final product (Gibis, 2016).

HCA are frequently observed in heated animal-derived foods such as beef, chicken, and fish due to the high creatine content required for their generation. A study done by Hasnol et al. (2014) on the effect of sugar marinating formulation on the formation of HCA in grilled chicken. The mean concentrations of MeIQ, PhIP, MeIQx, and norharman in the control samples were determined to be 16.4, 29.2, 18.9, and 17.9 ng/g, respectively, and were significantly higher ($p < 0.006$) than those with table sugar (9.03, 6.78, 12.9, and 8.41 ng/g), brown sugar (6.14, 6.74, 6.24, 11.5 ng/g), and honey (4.98, 9.87, 8.05, and 4.77 ng/g, respectively. In another study, Liao et al. (2010) studied on the effect of various cooking methods (pan-frying, deep-frying, charcoal grilling and roasting) on the formation of heterocyclic aromatic amines in chicken. The results indicated that overall concentration of nine HCA in unmarinated chicken grilled at surface temperatures of approximately 200 °C was 112 ng/g. The total concentrations of nine HCA in unmarinated chicken using method of pan-frying, deep-frying, and roasting were 27.4, 21.31 and 3.92 ng/g, respectively. It can be concluded that charcoal grilling and pan-frying are common methods for the preparation of poultry which yield a higher HCAs content than other cooking methods (Iqbal et al., 2016). Studies on the effect of cooking method towards marinated products are listed in Table 1.

Table 1a. Hetero-cyclic amine amounts (ng/g) in marinated product

Types of Meat	Cooking Method	Marinade Formulation	AαC	NorHarman	Harman	IQ
Chicken	Grilling	Honey	27.1± 0.39	4.77±0.59	8.99±0.99	4.09±0.96
Beef	Frying	13.5g buckwheat honey	-	-	-	-
Beef	Frying	13.5g clover honey	-	-	-	-
Chicken breast	Frying	13.5g buckwheat honey	-	-	-	-
Chicken breast	Frying	13.5g clover honey	-	-	-	-
Chicken breast	Grilling	Unmarinated	5.58±1.02	32.18±3.76	31.67±3.23	ND
Chicken breast	Grilling	Low concentration of organic acid	BDL	6.73 ± 1.09	BDL	22.2 ± 3.14
Chicken breast	Grilling	High concentration of organic acid	BDL	2.79 ± 1.09	4.05 ± 1.05	10.5 ± 2.51

Table 1b. Hetero-cyclic amine amounts (ng/g) in marinated product (Continued)

Types of Meat	MeIQ	IQx	MeIQx	4,8-DiMeIQx	7,8-DiMeIQx	PhIP	Reference
Chicken	4.98± 0.51	5.48± 0.81	8.05±0.59	-	-	9.87±0.35	Hasnol et al. (2014)
Beef	-	-	4.7±0.7	1.6±0.2	-	10.0±1.5	Shin and Ustunol (2006)
Beef	-	-	5.0±0.6	1.8±0.1	-	10.7±1.6	Shin and Ustunol (2006)
Chicken breast	-	-	3.6±	1.2±0.1	-	8.3±0.5	Shin and Ustunol (2006)
Chicken breast	-	-	4.1±	1.5±0.1	-	9.1±0.6	Shin and Ustunol (2006)
Chicken breast	-	-	1.83±0.86	1.05±0.40	-	31.06±4.53	Liao et al. (2010)
Chicken breast	5.77 ± 0.79	6.42±0.87	11.7 ± 0.97	23.1 ± 1.66	-	5.34 ± 0.97	Jinap et al. (2018)
Chicken breast	2.58 ± 0.97	3.39 ± 0.89	2.38 ± 0.99	11.3 ± 1.09	-	4.48 ± 1.02	Jinap et al. (2018)

PHYSICOCHEMICAL PROPERTIES OF MARINATED PRODUCTS

Chemical Properties of Marinated Products

Moisture Content Analysis

Moisture content is a surface textural characteristic that defines the perception of water absorbed or removed from food (Hudečková et al., 2013). It may influence the quality of final marinated product. Differences in temperature and moisture levels between the air and the product can cause evaporation of moisture at the surface of the product during cooking. As the product surface dries, the internal moisture will transport towards the product surface (Pathare & Roskilly, 2016). Thus, influences the texture and cooking loss of marinated (Jamali et al., 2016). A study done by Serdaroğlu et al. (2007) focused on the effects of marinating with citric acid solutions and grapefruit juice on cooking and eating quality of Turkey breast. Moisture contents of marinated uncooked samples ranged from 75.3 to 82.4% while the moisture contents ranged from 65.4 to 71.2% for samples cooked after marination. The highest moisture content was found in samples marinated with 50% grapefruit juice.

Therefore, it is concluded that cooking resulted in a moisture decrement in all treatment samples. Another study done by Smaoui et al. (2012) focused on the effect of sodium lactate and lactic acid combinations on the microbial, sensory, and chemical attributes of marinated chicken thigh. Five different formulations with different concentration of sodium lactate and lactic acid were used in this study. Among five different samples, the highest mean percentages generated for the moisture content was the formulation of 0.3% sodium lactate +0.03% lactic acid where that moisture content obtained was 72.6%. However, the formulation of 0.3% sodium lactate +0.03% lactic acid scores the lowest value (6.8) for the sensory attributes of texture.

pH Analysis

pH measurement is an acidity and basicity of aqueous solution, where most reactions occur naturally, depends on the concentration of hydronium (H_3O^+) and hydroxyl (OH^-) ions (Karastogianni et al., 2016). Lowering the pH of the meat muscle will contribute to an increase in the cooking loss of the meat, as it can lead to further protein denaturation and reduce the water holding capacity of meat products (Lu et al., 2018). This statement is supported by Yusop et al. (2010) in a study on the effect of low pH on the efficiency of marinade. The outcome showed that the pH varied

Table 2. Chemical properties of marinated products

Types of Meat	Cooking Method	Marinade Formulation	Chemical Properties			Reference
			Moisture Content	pH	DPPH Radical-Scavenging Activities (%)	
Chicken breast	Broiling	20% honey	-	5.99±0.07	-	Yücel et al. (2005)
Chicken breast	Broiling	30% honey	-	5.86±0.03	-	Yücel et al. (2005)
Chicken	Roast	lemon-pepper marinade+ 10% honey	84.73	6.55	-	Hashim et al. (1999)
Chicken	Roast	lemon-pepper marinade+ 20% honey	74.73	6.36	-	Hashim et al. (1999)
Chicken	Roast	lemon-pepper marinade+ 30% honey	64.73	6.22	-	Hashim et al. (1999)
Turkey's thigh	Grilling	Rosemary	-	5.8	-	Mielnik et al. (2008)
Beef	-	Nacl 2%	-	5.4	-	Sharedeh et al. (2015)
Beef	-	Nacl 0.9%	-	4.3	-	Sharedeh et al. (2015)
Longissimus dorsi muscles	-	2% NaCl (C1)	-	5.6±0.02	-	Fadiloğlu and Serdaroğlu (2018)
Longissimus dorsi muscles	-	2% NaCl+0.5 M Lactic Acid (LA1)	-	4.7±0.30	-	Fadiloğlu and Serdaroğlu (2018)
Longissimus dorsi muscles	-	2% NaCl+0.5 M Sodium Lactate (SL1)	-	5.4±0.06	-	Fadiloğlu and Serdaroğlu (2018)
Chicken breast fillet	Grilling	20% Chinese-style marinade solution	69.74 ± 0.90	4.2±0.08	-	Yusop et al. (2010)
Turkey breast meat	Roasting	1% STPP + 2% NaCl	81.50	6.29	-	Ergezer and, Gokce (2011)
Turkey breast meat	Roasting	0.5% lactic acid +3% NaCl	78.42	4.63	-	Ergezer and, Gokce (2011)
Chicken breast musle	Grilling	Acid Whey	-	5.81±0.03	-	Augustyńska-Prejsnar et al. (2019)

continued on following page

Table 2. Continued

Types of Meat	Cooking Method	Marinade Formulation	Chemical Properties			Reference
			Moisture Content	pH	DPPH Radical-Scavenging Activities (%)	
Mackerel	Barbequeimg	Salt	60.55 ± 8.56	5.98 ± 0.36	-	Oz and Kotan (2016)
Turkey breast meat	Steaming	100% grapefruit juice	70.3	4.0	-	Serdaroğlu et al. (2007)
Turkey breast meat	Steaming	0.2 M citric acid	70.3	3.9	-	Serdaroğlu et al. (2007)
Beef	Grilling	Gallic acid	-	-	7.5	Huiyuan Wang et al. (2017)
Beef	Grilling	Ferulic acid	-	-	12.5	Huiyuan Wang et al. (2017)
Beef	Grilling	Catechin	-	-	15	Huiyuan Wang et al. (2017)
Pork Loin	Charcoal Grilling	Pilsner	-	-	68	Viegas et al., (2014)
Pork Loin	Charcoal Grilling	Non-alcoholic Pilsner	-	-	36.5	Viegas et al. (2014)
Pork Loin	Charcoal Grilling	Black beer	-	-	29.5	Viegas et al. (2014)

from 3.0 to 3.6 contributed to a low water holding capacity (7.16- 8.44%). Thus, it is concluded that the importance of lower pH marinade in the determination of WHC of meat products is more commonly associated with the poor quality of WHC meat as described by higher cooking loss and juiciness. Moreover, Serdaroğlu et al. (2007) studied on the effects of marinating with citric acid solutions and grapefruit juice on Turkey breast. It is found that with increasing citric acid concentration (0.05, 0.1, and 0.2 M) and grapefruit concentration (50 and 100%), there was a clear decrease in muscle pH (from 5.8 to 3.9) and (from 4.1 to 4.0), respectively. Other findings on the variation in pH of marinade are shown in Table 2.

1,1-diphenyl-2-picrylhydrazyl (DPPH) Assay

DPPH assay is used to test the free radical scavenging ability of the antioxidant molecule and for evaluating the antioxidant properties of pure compounds (Viegas et al., 2014). This assay is widely used to calculate the antioxidant content of wheat grain and bran, vegetables, conjugated linoleic acids, spices, edible seed oils and flours in various solvent systems, including ethanol, aqueous acetone, methanol, aqueous alcohol, and benzene (Hussein et al., 2011; Mishra et al., 2012). Addition of different ingredients will affect the total phenolic and flavonoid content. Higher content of total phenolic and flavonoid content will increase the antioxidant content of the marinades. Thus, it will also increase the DPPH radical scavenging activities of the marinated products (Istrati et al., 2014). Hui Wang et al. (2017) studied on the effect of Catechin, gallic acid and ferulic acid towards antioxidants properties in charcoal-grilled meat.

The results indicated that formulation using catechin (21%) has the strongest scavenging activity prior to marinating compared to ferulic acid (12.5%) and gallic acid (7.5%) while after marinating, the strongest scavenging activity found was catechin (15%), followed by ferulic acid (12.5%) and gallic acid (7.5%). It is found that the different percentage of the scavenging for formulation using catechin was decrease and formulation using ferulic acid and gallic acid were increased. In addition, Viegas et al. (2015) studied on the influence of beer marinades on the reduction of carcinogenic heterocyclic aromatic amines in charcoal-grilled pork meat. Three different types of beers have been selected: black beer, non-alcoholic pilsner beer and pilsner beer, coded as BB, POB and PB. It is found that BB possessed the strongest 2,2-diphenyl-1-picrylhydrazyl (DPPH)-scavenging activity (68.0%) followed by POB (36.5%) and PB (29.5%).

Physical Properties of Marinated Product

Cooking Loss

Cooking loss is a mixture of liquid and soluble matter lost from the meat through heat treatment (Tänavots et al., 2018). It is a crucial factor in the meat industry as it determines the technological yield of the cooking process. Cooking loss will happen due to heat-induced protein denaturation during cooking. It is the most likely cause of water depletion during cooking, which allows less water to be entrapped within protein complexes retained by capillary forces (Lorenzo et al., 2015). Thermal protein denaturation results in improvements in protein-water interactions as well as geometrical improvements, such as transverse fibre shrinkage, due to altered lengths between muscle fibres and the development of pressure gradients (Margit Dall Aaslyng et al., 2003). Thus, alterations in the water-holding capacity of the meat will occurs and finally, the meat will lose water at high cooking temperatures (Zielbauer et al., 2016).

A study done by Hasnol et al. (2014) on the effect of three different marinating formulation on the formation of HCA in grilled chicken. The results indicated

Table 3. Physical properties of marinated products.

Types of Meat	Cooking Method	Marinade Formulation	Physical Properties			Reference
			Cooking Loss (%)	Drip Loss	Water-Holding Capacity/ Expressible Moisture (EM)	
Chicken breast	Broiling	0% honey	-	0.67±0.67	-	Yücel et al., (2005)
Chicken breast	Broiling	20% honey	-	1.66±0.28	-	Yücel et al., (2005)
Chicken breast	Broiling	30% honey	-	1.29±0.47	-	Yücel et al., (2005)
Chicken	Roast	lemon-pepper marinade+ 10% honey	23.98	-	-	Hashim et al., (1999)
Chicken	Roast	lemon-pepper marinade+ 20% honey	24.08	-	-	Hashim et al., (1999)
Chicken	Roast	lemon-pepper marinade+ 30% honey	24.14	-	-	Hashim et al., (1999)
Turkey's thigh	Grilling	Rosemary	27.2	-	-	Mielnik et al., (2008)
Longissimus dorsi muscles	-	2% NaCl (C1)	37.80±0.00	2.25 ± 0.34	-	Fadiloğlu and Serdaroğlu, (2018)
Longissimus dorsi muscles	-	2% NaCl+0.5 M Lactic Acid (LA1)	34.40±0.10	2.43±0.90	-	Fadiloğlu and Serdaroğlu, (2018)
Longissimus dorsi muscles	-	2% NaCl+0.5 M Sodium Lactate (SL1)	35.31±0.90	1.87±0.36	-	Fadiloğlu and Serdaroğlu, (2018)
Chicken breast fillet	Grilling	20% Chinese-style marinade solution	20.29 ± 2.78	-	8.23 ± 3.63	Yusop et al., (2010)
Chicken breast fillet	Roasting	Plum fiber	26.5	-	-	Jarvis et al., (2012)
Turkey breast meat	Roasting	1% STPP + 2% NaCl	41.69±5.55	5.1	-	Ergezer and Gokce (2011)
Turkey breast meat	Roasting	0.5% lactic acid +3% NaCl	47.21±2.73	0.5	-	Ergezer and Gokce (2011)
Turkey breast meat	Steaming	100% grapefruit juice	22.4	-	20.6	Serdaroğlu et al., (2007)
Turkey breast meat	Steaming	0.2 M citric acid	23.4	-	24.7	Serdaroğlu et al., (2007)

continued on following page

Table 3. Continued

Types of Meat	Cooking Method	Marinade Formulation	Physical Properties			Reference
			Cooking Loss (%)	Drip Loss	Water-Holding Capacity/ Expressible Moisture (EM)	
Chicken breast musle	Grilling	Acid Whey	-	-	28.73±2.86	Augustyńska-Prejsnar et al., (2019)
Chicken breast musle	Grilling	Lemon Juice	-	-	5.67±3.56	Augustyńska-Prejsnar et al., (2019)
Longissimus dorsi muscles	-	2% NaCl+0.5 M Lactic Acid (LA1)	34.40±0.10	2.43±0.90	-	Fadiloğlu and Serdaroğlu, (2018)

that the highest cooking loss was the control sample (41.2%) followed by chicken marinated with table sugar (38.5%), chicken marinated with brown sugar (38.4%) and chicken marinated with honey (35.2%) (see Table 3). There were no significant differences ($p > 0.05$) in cooking loss among the samples. Another study done by Yusop et al. (2010) on the effect of marinating time and low pH marinade on the physical properties of poultry meat. It is found that marinating for 60 minutes caused the highest of cooking loss (23.09%) while the lowest cooking loss were found when marinating time at 180 minutes. Besides, marinade pH of 3.2 caused highest cooking loss (24.04%) while marinade pH 4.2 resulted in the lowest cooking loss which was 20.29%.

Drip Loss

Drip loss is refers to fluids that contain primarily water and protein, which can be expelled from meat without mechanical force other than gravity (Fischer, 2007). The presence of expelled liquid in packaged meat makes the product unattractive to consumers and can induce microbial growth (Khiari et al., 2013). Drip loss can be affected due to the moisture content and water holding capacity of the meat (Jarvis et al., 2012). There are about 88–95% of the water in the muscle is retained intracellularly in the space between the actin and myosin filaments and the others is found between the myofibrils (Mir & Rafiq, 2017). High moisture content, reflecting the ability of muscle proteins to bind or immobilise a greater proportion of water molecules, thus decreasing in the drip loss (Traore et al., 2012). A study done by Yücel et al. (2005) on the drip loss of broiler breast meat using three different honey concentrations of 0, 20, and 30%. The highest drip loss obtained were formulation using 20% of honey followed by 30 and 0% of honey

(Table 3). There was no significant difference on the percentage of drip loss using different honey treatment level in breast meats. Besides, Gamage et al., (2016) had studied on the effect of marination methods (immersion, injection and tumbling) on the drip loss of broiler thigh meat. Among three methods, the highest drip loss obtained at 12 hours holding time was immersion methods (2.96%) followed by injection (2.57%) and tumbling (2.44%).

Water-holding Capacity

Water-holding capacity (WHC) is the ability of the meat to maintain its water when external force, such as heating is applied (Warner, 2023). The WHC measured as the percentage fraction of the total water content left after subtracting of the total juice loss (sum of the weight losses after centrifugation and cooking) (Lorenzo et al., 2015). Water holding capacity will increase due to the acidic pH of marinades. The mechanism of the tenderising action of acidic marinades is involving in the weakening of structures due to meat swelling and increased conversion of collagen to gelatine at low pH during cooking the meat (Klinhom et al., 2015). This statement is agreed by Serdaroğlu et al. (2007) in a study on the effects of marinating with citric acid solutions and grapefruit juice on water holding capacity of Turkey breast. The results indicated that the highest water holding capacity was formulation of 0.1 M citric acid (30.4%) with the pH of 3.2 while the lowest water holding capacity was using formulation of 100% grapefruit juice (20.6%) with the pH of 2.4.

However, this results in in contrary with a statement claimed by Yusop et al. (2010). They claimed that a significance of a lower pH marinade in the determination of WHC of meat products, most associated with the poor quality of WHC meat described by higher cooking loss, expressible moisture, and juiciness. Besides, WHC in meat products can increase by increasing the ionic strength. The ionic strength frees negatively charged sites on meat proteins, so that the proteins can bind more water (Yusop et al., 2011). In addition, more moisture content would remain in the meat resulting in high absorption of marinade, thereby increasing the WHC (Dhanda et al., 2002). Augustyńska-Prejsnar et al. (2019) has studied on the effects of marinating breast muscles of slaughter pheasants with acid whey, buttermilk, and lemon juice on quality parameters and product safety. The results indicated that high marinade absorption led to high water holding capacity. The highest marinade absorption (7.32%) was the formulation using buttermilk obtained WHC of 30.33%. The lowest marinade absorption (3.26%) was the formulation using lemon juice, obtained WHC of 25.67%.

Texture Profile Analysis (TPA)

Texture is the visual expression of the structural, mechanical, and surface properties of food products (Szczesniak, 2002). The texture of grilled product can be measured by using Texture Profile Analysis method (TPA) (Kruk et al., 2011). Its working theory uses a set of fixed force-correlation and shape-correlation features to describe the rheology of food, the tactile properties, and the impact of chewing time on these features. The instrument evaluates the texture properties of different foods using different model probes (Li, 2011). Meat tenderness can be affected by pH changes. When the pH rises and almost closer to neutral, the number of ionic charges rise between the muscle fibres, therefore increasing the amount of water transferred from free to immobilised. Next, the amount of water stored will be increased (Aguirre et al., 2018). Thus, high pH lead to higher WHC of meat and finally improves the texture of cooked meat (Yusop et al., 2011). Serdaroğlu et al. (2007) studied on the impacts of citric acid and grapefruit juice marinades on texture of Turkey breast.

The results found that marinade treatments significantly affected hardness and chewiness. Formulation of 100% grapefruit juice had similar chewiness with control group (2.3). Chewiness showed a similar trend to the hardness values. Besides, the result indicated that formulation of 0.2 M citric acid and 100% grapefruit juice marinade solutions increase the tenderness of turkey breast meat. The low in hardness values in 0.2 M citric acid and 100% grapefruit juice samples may be attributed to the effect of pH. The pH of turkey breast meat of 3.9 and 4.0 resulted in hardness of 1.8 and 2.2, respectively using Texture Profile Analysis. Other studies on the effect of marinade on the texture of marinated product are listed in Table 4.

Table 4. Texture Profile Analysis of marinated products

Types of Meat	Marinade Formulation	Texture Profile Analysis (TPA)					Reference
		Hardness (N)	Resilience	Cohesiveness	Springiness (mm)	Chewiness (MJ)	
Chicken breast musle	Acid Whey	17.18±2.80	0.18±0.04	0.23±0.05	1.41±0.28	5.40±1.86	Augustyńska-Prejsnar et al., (2019)
Chicken breast musle	Lemon Juice	18.29±2.02	0.19±0.04	0.29±0.04	1.76±0.42	9.53±2.56	Augustyńska-Prejsnar et al., (2019)
Turkey breast meat	100% grapefruit juice	2.2	1.7	-	-	2.3	Serdaroğlu et al., (2007)
Turkey breast meat	0.2 M citric acid	1.8	1.5	-	-	1.6	Serdaroğlu et al., (2007)

SENSORY PROPERTIES OF MARINATED PRODUCTS

Sensory assessment is used to enhance current products, flavour and colour recognition of products (Chilemba & Tanganyika, 2017). An effective test is one of the methods to evaluate the sensory properties of food. The affective test deliver subjective data on the acceptability, liking and preference of the products (Sharif et al., 2017). A 9-point hedonic rating scales may be used to assess the degree of satisfaction experienced with each sample (Sepahpour et al., 2018). Most common attributes has been studied for affective test on marinated product are flavour, juiciness, tenderness, odour, and overall acceptability (Hafez & Eissawy, 2018; Vlahova-Vangelova et al., 2016). Marination has been proven to be a good method to improve the sensory properties of meat products (Baygar et al., 2012).

There are several studies has been conducted on the sensory properties of marinated meat product using various ingredients. Yücel et al. (2005) has studied the sensory properties of honey treatment on some quality characteristics of broiler breast meat. Honey has been added at concentrations of 0, 20, and 30%. Taste panel results showed that there is no significant difference for tenderness, flavour, and acceptability among experimental groups. It is concluded that the flavour improvement would be because of fructose content in honey. In different study, Jarvis et al. (2012) focus on the sensory acceptability of dried plum as a substitute for phosphate in roasted chicken marinade. Five different formulations have been studied which are plum fibre, plum powder, plum concentrate, fibre/powder mix and sodium tripolyphosphate (STPP) as control. The results indicated that formulation with plum fibre/powder mix scores the highest mark (2.8, 2.5, 2.7%) for attributes of tenderness, juiciness, and flavour, respectively. It is concluded that the formulation of plum fiber/powder mix is suitable substitute for STPP as marinade in chicken breast. Another research has been reported on the sensory properties of marinated products are listed in Table 5.

FUTURE RESEARCH DIRECTIONS

Marinade has a strong popularity and a general interest in the properties of flavour and texture of the marinated products. Recent interest on the application of marinade has widely explored and brought to the introduction of honey marinade. Honey marinade plays an important role in reducing the formation of HCA and in enhancing the physicochemical properties of meat products. It can be used to satisfy the needs of customers who demand a good quality product with a natural ingredient in the marinade products. Work in this field has accelerated dramatically in recent years; however, there are still limited study focused on the powder marinades. Therefore,

Table 5. Sensory properties of marinated products

Types of Meat	Cooking Method	Marinade Formulation	Sensory Attributes						Reference
			1	2	3	4	5	6	
Chicken breast	Broiling	20% honey	7.36 0.48	-	6.45 0.42	-	-	6.73 0.2	Yücel et al., (2005)
Chicken breast	Broiling	30% honey	7.18 0.4	-	7.18 0.4	-	-	7.09 0.29	Yücel et al., (2005)
Turkey's thigh	Grilling	Rosemary	4.21	5.04	-	-	-	-	Mielnik et al., (2008)
Chicken breast fillet	Grilling	20% Chinese-style marinade solution	4.39	2.00	5.94	5.96	-	5.86	Yusop et al., (2010)
Chicken breast fillet	Roasting	Plum fibre/ powder mix	6.7	6.0	6.2	-	6.4	6.4	Jarvis et al., (2012)
Turkey breast meat	Steaming	100% grapefruit juice	7.3	6.8	5.2	-	6.3	-	Serdaroğlu et al., (2007)
Turkey breast meat	Steaming	0.2 M citric acid	6.5	6.9	2.9	-	2.5	-	Serdaroğlu et al., (2007)

Note: 1. Tenderness, 2. Juiciness, 3. Flavour , 4. Odour/ Aroma, 5. Appearance, 6. Overall acceptability.

it is suggested for future research to focus on nanotechnology of honey marinade. The use of nanotechnology will improve marinade performance by increasing the absorption of the marinade. Thus, the physicochemical properties of the marinated product may be enhanced, more flavourful and tender marinated products can also be produced.

CONCLUSION

In conclusion, this review proves that honey marinade has shown a potential in enhancing the physicochemical, and sensory properties of marinated products. The chemical properties such as moisture content, pH, and antioxidant activity as well as physical properties such as cooking loss, drip loss, water-holding capacity, and texture profile showed an improvement as compared to traditional marinade. Besides, sensory properties of honey marinades also give a good result however, the honey level in the marinades shown that there is no significant difference for tenderness, flavour, and acceptability of the honey marinated products. Moreover, based on the analysis and results discussed, the honey marinade showed the reduction levels of HCA in marinated products.

ACKNOWLEDGMENT

We would like to thank the reviewers for all their constructive comments and suggestions.

REFERENCES

Aaslyng, M. D, Lund, B. W., & Jensen, K. (2016). Inhibition of heterocyclic aromatic amines in pork chops using complex marinades with natural antioxidants. *Food and Nutrition Sciences*, *7*(14), 1315–1329. doi:10.4236/fns.2016.714120

Aaslyng, M. D., Bejerholm, C., Ertbjerg, P., Bertram, H. C., & Andersen, H. J. (2003). Cooking loss and juiciness of pork in relation to raw meat quality and cooking procedure. *Food Quality and Preference*, *14*(4), 277–288. doi:10.1016/S0950-3293(02)00086-1

Aguirre, M. E., Owens, C. M., Miller, R. K., & Alvarado, C. Z. (2018). Descriptive sensory and instrumental texture profile analysis of woody breast in marinated chicken. *Poultry Science*, *97*(4), 1456–1461. doi:10.3382/ps/pex428 PMID:29438548

Ahamed, M. M. E., Abdallah, A., Abdalaziz, A., Serag, E., & Atallah, A. E. H. (2017). Some physiochemical properties of Acacia honey from different altitudes of the Asir Region in Southern Saudi Arabia. *Czech Journal of Food Sciences*, *35*(4), 321–327. doi:10.17221/428/2016-CJFS

Ahmed, S., & Othman, N. H. (2013). Review of the medicinal effects of tualang honey and a comparison with manuka honey. *The Malaysian Journal of Medical Sciences*, *20*(3), 6–13. PMID:23966819

Ahmed, S., & Othman, N. H. (2017). The anti-cancer effects of Tualang honey in modulating breast carcinogenesis: An experimental animal study. *BMC Complementary and Alternative Medicine*, *17*(1), 1–11. doi:10.118612906-017-1721-4 PMID:28399853

Alaejos, M. S., & Afonso, A. M. (2011). Factors that affect the content of heterocyclic aromatic amines in foods. *Comprehensive Reviews in Food Science and Food Safety*, *10*(2), 52–108. doi:10.1111/j.1541-4337.2010.00141.x

Ali, A., Waly, M. I., & Devarajan, S. (2019). *Impact of processing meat on the formation of heterocyclic amines and risk of cancer. In Biogenic Amines in Food: Analysis, Occurrence and Toxicity* (1st ed.). The Royal Society of Chemistry.

Alvarado, C., & McKee, S. (2007). Marination to improve functional properties and safety of poultry meat. *Journal of Applied Poultry Research, 16*(1), 113–120. doi:10.1093/japr/16.1.113

Arcanjo, N. M. O., Ventanas, S., González-Mohíno, A., Madruga, M. S., & Estévez, M. (2019). Benefits of wine-based marination of strip steaks prior to roasting: Inhibition of protein oxidation and impact on sensory properties. *Journal of the Science of Food and Agriculture, 99*(3), 1108–1116. doi:10.1002/jsfa.9278 PMID:30047154

Augustyńska-Prejsnar, A., Ormian, M., Hanus, P., Kluz, M., Sokołowicz, Z., & Rudy, M. (2019). Effects of marinating breast muscles of slaughter pheasants with acid whey, buttermilk, and lemon juice on quality parameters and product safety. *Journal of Food Quality, 2019*, 1–8. doi:10.1155/2019/5313496

Ávila, S., Lazzarotto, M., Hornung, P. S., Teixeira, G. L., Ito, V. C., Bellettini, M. B., Beux, M. R., & Ribani, R. H. (2019). Influence of stingless bee genus (Scaptotrigona and Melipona) on the mineral content, physicochemical and microbiological properties of honey. *Journal of Food Science and Technology, 56*(10), 4742–4748. doi:10.100713197-019-03939-8 PMID:31686706

Azonwade, F. E., Paraïso, A., Agbangnan Dossa, C. P., Dougnon, V. T., N'tcha, C., Mousse, W., & Baba-Moussa, L. (2018). Physicochemical characteristics and microbiological quality of honey produced in Benin. *Journal of Food Quality, 56*, 4742–4648. doi:10.1155/2018/1896057

Babikova, J., Hoeche, U., Boyd, J., & Noci, F. (2020). Nutritional, physical, microbiological, and sensory properties of marinated Irish sprat. *International Journal of Gastronomy and Food Science, 22*, 100277. doi:10.1016/j.ijgfs.2020.100277

Batt, P. J., & Liu, A. (2012). Consumer behaviour towards honey products in Western Australia. *British Food Journal, 114*(2), 285–297. doi:10.1108/00070701211202449

Baygar, T., Alparslan, Y., & Kaplan, M. (2012). Determination of changes in chemical and sensory quality of sea bass marinades stored at + 4 (±1) C in marinating solution. *CYTA: Journal of Food, 10*(3), 196–200. doi:10.1080/19476337.2011.614016

Behera, S. S., Madathil, D., Verma, S. K., & Pathak, N. (2020). Seafood marination-A review. *International Archive of Applied Sciences and Technology, 11*(3), 165–168.

Birk, T., Grønlund, A. C., Christensen, B. B., Knøchel, S., Lohse, K., & Rosenquist, H. (2010). Effect of organic acids and marination ingredients on the survival of Campylobacter jejuni on meat. *Journal of Food Protection, 73*(2), 258–265. doi:10.4315/0362-028X-73.2.258 PMID:20132670

Bodor, Z., Koncz, F. A., Zinia Zaukuu, J.-L., Kertész, I., Gillay, Z., Kaszab, T., Kovács, Z., & Benedek, C. (2017). Effect of heat treatment on chemical and sensory properties of honeys. *Animal Welfare, Etológia És Tartástechnológia, 13*(2), 39–48.

Budin, S. B., Jubaidi, F. F., Azam, S. N. F. M. N., Yusof, N. L. M., Taib, I. S., & Mohamed, J. (2017). Kelulut honey supplementation prevents sperm and testicular oxidative damage in streptozotocin-induced diabetic rats. *Jurnal Teknologi, 79*(3), 90–95. doi:10.11113/jt.v79.9674

Casey, B. J., Jones, R. M., Levita, L., Libby, V., Pattwell, S. S., Ruberry, E. J., Soliman, F., & Somerville, L. H. (2010). The storm and stress of adolescence: Insights from human imaging and mouse genetics. *Developmental Psychobiology: The Journal of the International Society for Developmental Psychobiology, 52*(3), 225–235. doi:10.1002/dev.20447 PMID:20222060

Chan, B. K., Haron, H., Talib, R. A., & Subramaniam, P. (2017). Physical properties, antioxidant content and anti-oxidative activities of Malaysian stingless kelulut (Trigona spp.) honey. *J. Agric. Sci, 9*(13), 32–40.

Chen, X., Jia, W., Zhu, L., Mao, L., & Zhang, Y. (2020). Recent advances in heterocyclic aromatic amines: An update on food safety and hazardous control from food processing to dietary intake. *Comprehensive Reviews in Food Science and Food Safety, 19*(1), 124–148. doi:10.1111/1541-4337.12511 PMID:33319523

Cheng, K., Chen, F., & Wang, M. (2006). Heterocyclic amines: Chemistry and health. *Molecular Nutrition & Food Research, 50*(12), 1150–1170. doi:10.1002/mnfr.200600086 PMID:17131456

Cheng, K., Chen, F., & Wang, M. (2007). Inhibitory activities of dietary phenolic compounds on heterocyclic amine formation in both chemical model system and beef patties. *Molecular Nutrition & Food Research, 51*(8), 969–976. doi:10.1002/mnfr.200700032 PMID:17628877

Cheung, Y., Meenu, M., Yu, X., & Xu, B. (2019). Phenolic acids and flavonoids profiles of commercial honey from different floral sources and geographic sources. *International Journal of Food Properties, 22*(1), 290–308. doi:10.1080/10942912.2019.1579835

Chilemba, T., & Tanganyika, J. (2017). The effects of mode of transport on sensory characteristics in indigenous Malawian chickens. *Int J Avian & Wildlife Biol, 2*(4), 115–118.

Chong, K. Y., Chin, N. L., & Yusof, Y. A. (2017). Thermosonication and optimization of stingless bee honey processing. *Food Science & Technology International, 23*(7), 608–622. doi:10.1177/1082013217713331 PMID:28614964

Chu, Y., Huang, C., Xie, X., Tan, B., Kamal, S., & Xiong, X. (2018). Multilayer hybrid deep-learning method for waste classification and recycling. *Computational Intelligence and Neuroscience*, *2018*, 1–9. doi:10.1155/2018/5060857 PMID:30515197

Cianciosi, D., Forbes-Hernández, T. Y., Afrin, S., Gasparrini, M., Reboredo-Rodriguez, P., Manna, P. P., Zhang, J., Bravo Lamas, L., Martínez Flórez, S., Agudo Toyos, P., Quiles, J., Giampieri, F., & Battino, M. (2018). Phenolic compounds in honey and their associated health benefits: A review. *Molecules (Basel, Switzerland)*, *23*(9), 1–20. doi:10.3390/molecules23092322 PMID:30208664

Czipa, N., Phillips, C. J. C., & Kovács, B. (2019). Composition of acacia honeys following processing, storage and adulteration. *Journal of Food Science and Technology*, *56*(3), 1245–1255. doi:10.100713197-019-03587-y PMID:30956304

Daly, M., Halpin, E., Dawson, P., & Acton, J. (2013). Properties of injection-marinated chicken breasts. *XXI European Symposium on the Quality of Poultry Meat*, 15–19.

Dan, P. N. S. M., Omar, S., & Ismail, W. I. W. (2018). Physicochemical analysis of several natural malaysian honeys and adulterated honey. *IOP Conference Series. Materials Science and Engineering*, *440*(1), 12049. doi:10.1088/1757-899X/440/1/012049

De-Melo, A. A. M., de Almeida-Muradian, L. B., Sancho, M. T., & Pascual-Maté, A. (2018). Composición y propiedades de la miel de Apis mellifera: Una revisión. *Journal of Apicultural Research*, *57*(1), 5–37.

De Silva, T., Uneri, A., Ketcha, M. D., Reaungamornrat, S., Kleinszig, G., Vogt, S., Aygun, N., Lo, S. F., Wolinsky, J. P., & Siewerdsen, J. H. (2016). 3D–2D image registration for target localization in spine surgery: Investigation of similarity metrics providing robustness to content mismatch. *Physics in Medicine and Biology*, *61*(8), 3009–3025. doi:10.1088/0031-9155/61/8/3009 PMID:26992245

Dhanda, J. S., Pegg, R. B., Janz, J. A. M., Aalhus, J. L., & Shand, P. J. (2002). Palatability of bison semimembranosus and effects of marination. *Meat Science*, *62*(1), 19–26. doi:10.1016/S0309-1740(01)00222-4 PMID:22061187

Erge, A., Cin, K., & Şeker, E. (2018). The use of plum and apple juice at chicken meat marination. *GIDA-Journal of Food*, *43*(6), 1040–1052. doi:10.15237/gida.GD18063

Ergezer, H., & Gokce, R. (2011). Comparison of marinating with two different types of marinade on some quality and sensory characteristics of Turkey breast meat. *Journal of Animal and Veterinary Advances*, *10*(1), 60–67. doi:10.3923/javaa.2011.60.67

Eshete, Y., & Eshete, T. (2019). A review on the effect of processing temperature and time duration on commercial honey quality. *Madridge Journal of Food Technology*, *4*(1), 158–162. doi:10.18689/mjft-1000124

Eteraf-Oskouei, T., & Najafi, M. (2013). Traditional and modern uses of natural honey in human diseases: A review. *Iranian Journal of Basic Medical Sciences.*, *16*(6), 731–742. PMID:23997898

Fadiloğlu, E. E., & Serdaroğlu, M. (2018). Effects of pre and post-rigor marinade injection on some quality parameters of Longissimus dorsi muscles. *Han-gug Chugsan Sigpum Hag-hoeji*, *38*(2), 325–337. doi:10.5851/kosfa.2018.38.2.325 PMID:29805282

Fatima, I. J., AB, M. H., Salwani, I., & Lavaniya, M. (2018). Physicochemical characteristics of malaysian stingless bee honey from trigona species. *IIUM Medical Journal Malaysia, 17*(1), 187–191. doi:10.31436/imjm.v17i1.1030

Fischer, K. (2007). Drip loss in pork: Influencing factors and relation to further meat quality traits. *Journal of Animal Breeding and Genetics*, *124*(s1), 12–18. doi:10.1111/j.1439-0388.2007.00682.x PMID:17988246

Fu, Z., Deming, S. L., Fair, A. M., Shrubsole, M. J., Wujcik, D. M., Shu, X.-O., Kelley, M., & Zheng, W. (2011). Well-done meat intake and meat-derived mutagen exposures in relation to breast cancer risk: The Nashville Breast Health Study. *Breast Cancer Research and Treatment*, *129*(3), 919–928. doi:10.100710549-011-1538-7 PMID:21537933

Gamage, H., Mutucumarana, R. K., & Andrew, M. S. (2017). The Effect of Marination Method and Holding Time on Physicochemical and Sensory Properties of Broiler Meat. *Journal of Agricultural Sciences*, *12*(3), 172–184. doi:10.4038/jas.v12i3.8264

Gao, T., Li, J., Zhang, L., Jiang, Y., Yin, M., Liu, Y., Gao, F., & Zhou, G. (2015). Effect of different tumbling marination methods and time on the water status and protein properties of prepared pork chops. *Asian-Australasian Journal of Animal Sciences*, *28*(7), 1020–1027. doi:10.5713/ajas.14.0918 PMID:26104408

Gibis, M. (2016). Heterocyclic aromatic amines in cooked meat products: Causes, formation, occurrence, and risk assessment. *Comprehensive Reviews in Food Science and Food Safety*, *15*(2), 269–302. doi:10.1111/1541-4337.12186 PMID:33371602

Goli, T., Ricci, J., Bohuon, P., Marchesseau, S., & Collignan, A. (2014). Influence of sodium chloride and pH during acidic marination on water retention and mechanical properties of turkey breast meat. *Meat Science*, *96*(3), 1133–1140. doi:10.1016/j.meatsci.2013.10.031 PMID:24334031

Hafez, A., & Eissawy, M. M. (2018). Effect of banana peel extract on sensory and bacteriological quality of marinated beef. *Archives of Nutrition and Public Health*, *1*(1), 1–11.

Hashim, I. B., McWatters, K. H., & Hung, Y. (1999). Marination method and honey level affect physical and sensory characteristics of roasted chicken. *Journal of Food Science*, *64*(1), 163–166. doi:10.1111/j.1365-2621.1999.tb09883.x

Hasnol, N. D. S., Jinap, S., & Sanny, M. (2014). Effect of different types of sugars in a marinating formulation on the formation of heterocyclic amines in grilled chicken. *Food Chemistry*, *145*, 514–521. doi:10.1016/j.foodchem.2013.08.086 PMID:24128508

Hudečková, M., Vojtíšková, P., & Kráčmar, S. (2013). The influence of different conditions on the textural properties of meat during grilling. *Journal of Microbiology, Biotechnology and Food Sciences*, *10*(1), 1225–1230.

Hussein, S. Z., Yusoff, K. M., Makpol, S., & Yusof, Y. A. M. (2011). Antioxidant capacities and total phenolic contents increase with gamma irradiation in two types of Malaysian honey. *Molecules (Basel, Switzerland)*, *16*(8), 6378–6395. doi:10.3390/molecules16066378 PMID:21796076

Ibrahim, H. M., Amin, R. A., & Ghanaym, H. R. M. (2018). Effect of marination on Vibrio parahaemolyticus in tilapia fillets. *Benha Veterinary Medical Journal*, *34*(2), 234–245. doi:10.21608/bvmj.2018.29434

Iftikhar, F., Arshad, M., Rasheed, F., Amraiz, D., Anwar, P., & Gulfraz, M. (2010). Effects of acacia honey on wound healing in various rat models. *Phytotherapy Research: An International Journal Devoted to Pharmacological and Toxicological Evaluation of Natural Product Derivatives*, *24*(4), 583–586. doi:10.1002/ptr.2990 PMID:19813239

Inguglia, E. S., Burgess, C. M., Kerry, J. P., & Tiwari, B. K. (2019). Ultrasound-assisted marination: Role of frequencies and treatment time on the quality of sodium-reduced poultry meat. *Foods*, *8*(10), 1–11. doi:10.3390/foods8100473 PMID:31614455

Iqbal, S. Z., Talib, N. H., & Hasnol, N. D. S. (2016). Heterocyclic aromatic amines in deep fried lamb meat: The influence of spices marination and sensory quality. *Journal of Food Science and Technology*, *53*(3), 1411–1417. doi:10.100713197-015-2137-0 PMID:27570265

Istrati, D., Constantin, O., Vizireanu, C., & Dinica, R. M. (2014). The study of antioxidant and antimicrobial activity of extracts for meat marinades. *Romanian Biotechnological Letters*, *19*(5), 9687–9698.

Istrati, D., Simion Ciuciu, A., Vizireanu, C., Ionescu, A., & Carballo, J. (2015). Impact of Spices and Wine-Based Marinades on Tenderness, Fragmentation of Myofibrillar Proteins and Color Stability in Bovine B iceps Femoris Muscle. *Journal of Texture Studies*, *46*(6), 455–466. doi:10.1111/jtxs.12144

Jamali, M. A., Zhang, Y., Teng, H., Li, S., Wang, F., & Peng, Z. (2016). Inhibitory effect of rosa rugosa tea extract on the formation of heterocyclic amines in meat patties at different temperatures. *Molecules (Basel, Switzerland)*, *21*(2), 1–14. doi:10.3390/molecules21020173 PMID:26840288

Jambrak, A. R. (2017). Physical properties of sonicated products: a new era for novel ingredients. In Ultrasound: Advances for Food Processing and Preservation (pp. 237–265). Elsevier.

Jarvis, N., Clement, A. R., O'Bryan, C. A., Babu, D., Crandall, P. G., Owens, C. M., Meullenet, J., & Ricke, S. C. (2012). Dried plum products as a substitute for phosphate in chicken marinade. *Journal of Food Science*, *77*(6), S253–S257. doi:10.1111/j.1750-3841.2012.02737.x PMID:22671532

Jinap, S., Hasnol, N. D. S., Sanny, M., & Jahurul, M. H. A. (2018). Effect of organic acid ingredients in marinades containing different types of sugar on the formation of heterocyclic amines in grilled chicken. *Food Control*, *84*, 478–484. doi:10.1016/j.foodcont.2017.08.025

Jinap, S., Iqbal, S. Z., & Selvam, R. M. P. (2015). Effect of selected local spices marinades on the reduction of heterocyclic amines in grilled beef (satay). *Lebensmittel-Wissenschaft + Technologie*, *63*(2), 919–926. doi:10.1016/j.lwt.2015.04.047

Kameník, J., Saláková, A., Vyskočilová, V., Pechová, A., & Haruštiaková, D. (2017). Salt, sodium chloride or sodium? Content and relationship with chemical, instrumental and sensory attributes in cooked meat products. *Meat Science*, *131*, 196–202. doi:10.1016/j.meatsci.2017.05.010 PMID:28551463

Karabagias, I. K., Badeka, A., Kontakos, S., Karabournioti, S., & Kontominas, M. G. (2014). Characterisation and classification of Greek pine honeys according to their geographical origin based on volatiles, physicochemical parameters and chemometrics. *Food Chemistry*, *146*, 548–557. doi:10.1016/j.foodchem.2013.09.105 PMID:24176380

Karam, L., Roustom, R., Abiad, M. G., El-Obeid, T., & Savvaidis, I. N. (2019). Combined effects of thymol, carvacrol and packaging on the shelf-life of marinated chicken. *International Journal of Food Microbiology*, *291*, 42–47. doi:10.1016/j.ijfoodmicro.2018.11.008 PMID:30445284

Karastogianni, S., Girousi, S., & Sotiropoulos, S. (2016). pH: principles and measurement. Encyclopedia of Food and Health, 4, 333–338.

Kataoka, H., Miyake, M., Saito, K., & Mitani, K. (2012). Formation of heterocyclic amine–amino acid adducts by heating in a model system. *Food Chemistry*, *130*(3), 725–729. doi:10.1016/j.foodchem.2011.07.094

Kek, S. P., Chin, N. L., Yusof, Y. A., Tan, S. W., & Chua, L. S. (2014). Total phenolic contents and colour intensity of Malaysian honeys from the Apis spp. and Trigona spp. bees. *Agriculture and Agricultural Science Procedia*, *2*, 150–155. doi:10.1016/j. aaspro.2014.11.022

Kek, S. P., Chin, N. L., Yusof, Y. A., Tan, S. W., & Chua, L. S. (2017). Classification of entomological origin of honey based on its physicochemical and antioxidant properties. *International Journal of Food Properties,* *20*(sup3), S2723–S2738.

Khalil, M., Tanvir, E. M., Afroz, R., Sulaiman, S. A., & Gan, S. H. (2015). Cardioprotective effects of tualang honey: amelioration of cholesterol and cardiac enzymes levels. *BioMed Research International, 2015*(8). doi:10.1155/2015/286051

Khiari, Z., Omana, D. A., Pietrasik, Z., & Betti, M. (2013). Evaluation of poultry protein isolate as a food ingredient: Physicochemical properties and sensory characteristics of marinated chicken breasts. *Journal of Food Science*, *78*(7), S1069–S1075. doi:10.1111/1750-3841.12167 PMID:23772877

Kim, H.-Y., Kim, K.-J., Lee, J.-W., Kim, G.-W., Choe, J.-H., Kim, H.-W., Yoon, Y., & Kim, C.-J. (2015). Quality characteristics of marinated chicken breast as influenced by the methods of mechanical processing. *Han-gug Chugsan Sigpum Hag-hoeji*, *35*(1), 101–107. doi:10.5851/kosfa.2015.35.1.101 PMID:26761806

Kim, S.-Y., Song, D.-H., Ham, Y.-K., Choi, Y.-S., Choi, J.-H., & Kim, H.-W. (2019). Efficacy of tumbling in soy sauce marination of pork loins: Effects of tumbling time and temperature. *Journal of Food Science and Technology*, *56*(12), 5282–5288. doi:10.100713197-019-03997-y PMID:31749475

Klinhom, P., Klinhom, J., Senapa, J., & Methawiwat, S. (2015). Improving the quality of citric acid and calcium chloride marinated culled cow meat. *International Food Research Journal*, *22*(4), 1410–1416.

Kruk, Z. A., Yun, H., Rutley, D. L., Lee, E. J., Kim, Y. J., & Jo, C. (2011). The effect of high pressure on microbial population, meat quality and sensory characteristics of chicken breast fillet. *Food Control*, *22*(1), 6–12. doi:10.1016/j.foodcont.2010.06.003

Kumar, Y., Yadav, D. N., Ahmad, T., & Narsaiah, K. (2015). Recent trends in the use of natural antioxidants for meat and meat products. *Comprehensive Reviews in Food Science and Food Safety*, *14*(6), 796–812. doi:10.1111/1541-4337.12156

Ledward, D. A. (2003). Meat: preservation. In Encyclopedia of Food Sciences and Nutrition (pp. 3772–3777). Academic Press. doi:10.1016/B0-12-227055-X/00752-5

Lee, Y. S., Zhekov, Z. G., Owens, C. M., Kim, M., & Meullenet, J. F. (2012). Effects of partial and complete replacement of sodium chloride with potassium chloride on the texture, flavor and water-holding capacity of marinated broiler breast fillets. *Journal of Texture Studies*, *43*(2), 124–132. doi:10.1111/j.1745-4603.2011.00322.x

Li, H., & Gänzle, M. (2016). Effect of hydrostatic pressure and antimicrobials on survival of Listeria monocytogenes and enterohaemorrhagic Escherichia coli in beef. *Innovative Food Science & Emerging Technologies*, *38*, 321–327. doi:10.1016/j.ifset.2016.05.003

Li, X., Sun, Y., Pan, D., Wang, Y., & Cao, J. (2017). The effect of CaCl2 marination on the tenderizing pathway of goose meat during conditioning. *Food Research International*, *102*, 487–492. doi:10.1016/j.foodres.2017.09.014 PMID:29195976

Li, X. M. (2011). Correlation analysis between measured values of the texture analyzer and scale values of sensory evaluation for food hardness. *Advanced Materials Research*, *183*, 882–886. doi:10.4028/www.scientific.net/AMR.183-185.882

Liao, G. Z., Wang, G. Y., Xu, X. L., & Zhou, G. H. (2010). Effect of cooking methods on the formation of heterocyclic aromatic amines in chicken and duck breast. *Meat Science*, *85*(1), 149–154. doi:10.1016/j.meatsci.2009.12.018 PMID:20374878

Liza A-Rahaman, N., Suan Chua, L., Roji Sarmidi, M., & Aziz, R. (2013). Physicochemical and radical scavenging activities of honey samples from Malaysia. *Agricultural Sciences*, *4*(5), 46–51. doi:10.4236/as.2013.45B009

Lorenzo, J. M., Cittadini, A., Munekata, P. E., & Domínguez, R. (2015). Physicochemical properties of foal meat as affected by cooking methods. *Meat Science*, *108*, 50–54. doi:10.1016/j.meatsci.2015.05.021 PMID:26042921

Lu, F., Kuhnle, G. K., & Cheng, Q. (2018). The effect of common spices and meat type on the formation of heterocyclic amines and polycyclic aromatic hydrocarbons in deep-fried meatballs. *Food Control*, *92*, 399–411. doi:10.1016/j.foodcont.2018.05.018

Lytou, A. E., Nychas, G.-J. E., & Panagou, E. Z. (2018). Effect of pomegranate based marinades on the microbiological, chemical and sensory quality of chicken meat: A metabolomics approach. *International Journal of Food Microbiology*, *267*, 42–53. doi:10.1016/j.ijfoodmicro.2017.12.023 PMID:29288907

Manful, C. F., Vidal, N. P., Pham, T. H., Nadeem, M., Wheeler, E., Hamilton, M. C., Doody, K. M., & Thomas, R. H. (2020). Unfiltered beer based marinades reduced exposure to carcinogens and suppressed conjugated fatty acid oxidation in grilled meats. *Food Control, 111*(107040), 1–15. doi:10.1016/j.foodcont.2019.107040

Mielnik, M. B., Sem, S., Egelandsdal, B., & Skrede, G. (2008). By-products from herbs essential oil production as ingredient in marinade for turkey thighs. *Lebensmittel-Wissenschaft + Technologie, 41*(1), 93–100. doi:10.1016/j.lwt.2007.01.014

Mir, N. A., & Rafiq, A. (2017). Faneshwar Kumar, Vijay Singh & Vivek Shukla. *Journal of Food Science and Technology, 54*, 2997–3009. doi:10.100713197-017-2789-z PMID:28974784

Mishra, K., Ojha, H., & Chaudhury, N. K. (2012). Estimation of antiradical properties of antioxidants using DPPH assay: A critical review and results. *Food Chemistry, 130*(4), 1036–1043. doi:10.1016/j.foodchem.2011.07.127

Mohammed, N. S., Mansour, E. H., Osheba, A. S., & Hassan, A. A. (2017). Effect of acidic marination on the quality characteristics of spent hen kobeba during frozen storage. *Arab Universities Journal of Agricultural Sciences, 25*(1), 157–167. doi:10.21608/ajs.2017.13399

Ngoi, V. (2016). *Effect of processing treatment on antioxidant, physicochemical and enzymatic properties of honey (Trigona spp.).* UTAR.

Nordin, A., Sainik, N. Q. A. V., Chowdhury, S. R., Saim, A., & Idrus, R. B. H. (2018). Physicochemical properties of stingless bee honey from around the globe: A comprehensive review. *Journal of Food Composition and Analysis, 73*, 91–102. doi:10.1016/j.jfca.2018.06.002

O'Neill, C. M., Cruz-Romero, M. C., Duffy, G., & Kerry, J. P. (2019). Improving marinade absorption and shelf life of vacuum packed marinated pork chops through the application of high pressure processing as a hurdle. *Food Packaging and Shelf Life, 21*(100350), 1–11. doi:10.1016/j.fpsl.2019.100350

Ouali, A., Gagaoua, M., Boudida, Y., Becila, S., Boudjellal, A., Herrera-Mendez, C. H., & Sentandreu, M. A. (2013). Biomarkers of meat tenderness: Present knowledge and perspectives in regards to our current understanding of the mechanisms involved. *Meat Science, 95*(4), 854–870. doi:10.1016/j.meatsci.2013.05.010 PMID:23790743

Oyeyinka, S. A., Alabi-Ogundepo, T., Babayeju, A. A., & Joseph, J. K. (2019). Consumer awareness, proximate composition, and sensory properties of processed African giant rat (Cricetomys gambianus) thigh meat. *Journal of the Saudi Society of Agricultural Sciences, 18*(4), 385–388. doi:10.1016/j.jssas.2017.12.006

Oz, F., & Cakmak, I. H. (2016). The effects of conjugated linoleic acid usage in meatball production on the formation of heterocyclic aromatic amines. *Lebensmittel-Wissenschaft + Technologie*, *65*, 1031–1037. doi:10.1016/j.lwt.2015.09.040

Oz, F., & Kotan, G. (2016). Effects of different cooking methods and fat levels on the formation of heterocyclic aromatic amines in various fishes. *Food Control*, *67*, 216–224. doi:10.1016/j.foodcont.2016.03.013

Ozogul, Y., & Balikci, E. (2013). Effect of various processing methods on quality of mackerel (Scomber scombrus). *Food and Bioprocess Technology*, *6*(4), 1091–1098. doi:10.100711947-011-0641-4

Palang, E. Y. (2004). *The role of ingredients and processing conditions on marinade penetration, retention and color defects in cooked marinated chicken breast meat.* University of Georgia.

Pasias, I. N., Kiriakou, I. K., & Proestos, C. (2017). HMF and diastase activity in honeys: A fully validated approach and a chemometric analysis for identification of honey freshness and adulteration. *Food Chemistry*, *229*, 425–431. doi:10.1016/j. foodchem.2017.02.084 PMID:28372195

Pathare, P. B., & Roskilly, A. P. (2016). Quality and energy evaluation in meat cooking. *Food Engineering Reviews*, *8*(4), 435–447. doi:10.100712393-016-9143-5

Pérez-Juan, M., Kondjoyan, A., Picouet, P., & Realini, C. E. (2012). Effect of marination and microwave heating on the quality of Semimembranosus and Semitendinosus muscles from Friesian mature cows. *Meat Science*, *92*(2), 107–114. doi:10.1016/j.meatsci.2012.04.020 PMID:22578362

Pfau, W., & Skog, K. (2004). Exposure to β-carbolines norharman and harman. *Journal of Chromatography. B, Analytical Technologies in the Biomedical and Life Sciences*, *802*(1), 115–126. doi:10.1016/j.jchromb.2003.10.044 PMID:15036003

Pimentel-González, D. J., Jiménez-Alvarado, R., Hernández-Fuentes, A. D., Figueira, A. C., Suarez-Vargas, A., & Campos-Montiel, R. G. (2016). Potentiation of bioactive compounds and antioxidant activity in artisanal honeys using specific heat treatments. *Journal of Food Biochemistry*, *40*(1), 47–52. doi:10.1111/jfbc.12186

Pleva, D., Lányi, K., Monori, K. D., & Laczay, P. (2020). Heterocyclic amine formation in grilled chicken depending on body parts and treatment conditions. *Molecules (Basel, Switzerland)*, *25*(7), 1547. doi:10.3390/molecules25071547 PMID:32231032

Pooona, J., Singh, P., & Prabhakaran, P. (2019). Effect of kiwifruit juice and tumbling on tenderness and lipid oxidation in chicken nuggets. *Nutrition & Food Science*, *50*(1), 74–83. doi:10.1108/NFS-12-2018-0352

Prica, N., Baloš, M. Ž., Jakšić, S., Mihaljev, Ž., Kartalović, B., Babić, J., & Savić, S. (2014). Moisture and acidity as indicators of the quality of honey originating from Vojvodina region. *Arhiv Veterinarske Medicine*, *7*(2), 99–109. doi:10.46784/e-avm.v7i2.135

Puangsombat, K. (2010). *Formation and inhibition of heterocyclic amines in meat products*. Kansas State University.

Quelhas, I., Petisca, C., Viegas, O., Melo, A., Pinho, O., & Ferreira, I. (2010). Effect of green tea marinades on the formation of heterocyclic aromatic amines and sensory quality of pan-fried beef. *Food Chemistry*, *122*(1), 98–104. doi:10.1016/j.foodchem.2010.02.022

Rahman, M. M., Khalil, D. M. I., & Gan, S. H. (2014). Neurological effects of honey: Current and future prospects. *Evidence-Based Complementary and Alternative Medicine*, *10*(958721), 1–13. doi:10.1155/2014/958721

Rao, P. V., Krishnan, K. T., Salleh, N., & Gan, S. H. (2016). Biological and therapeutic effects of honey produced by honey bees and stingless bees: A comparative review. *Revista Brasileira de Farmacognosia*, *26*(5), 657–664. doi:10.1016/j.bjp.2016.01.012

Sahar, A., Khan, M. I., & Khan, M. A. (2016). Heterocyclic Amines. In *Food Safety* (pp. 89–111). Springer. doi:10.1007/978-3-319-39253-0_5

Sajid, M., Yasmin, T., Asad, F., & Qamer, S. (2019). 66. Changes in HMF content and diastase activity in honey after heating treatment. *Pure and Applied Biology*, *8*(2), 1668–1674. doi:10.19045/bspab.2019.80109

Sak-Bosnar, M., & Sakač, N. (2012). Direct potentiometric determination of diastase activity in honey. *Food Chemistry*, *135*(2), 827–831. doi:10.1016/j.foodchem.2012.05.006 PMID:22868165

Samira, N. (2016). The effect of heat treatment on the quality of Algerian honey. *Researcher*, *8*(9), 1–6.

Scalone, G. L. L., Lamichhane, P., Cucu, T., De Kimpe, N., & De Meulenaer, B. (2019). Impact of different enzymatic hydrolysates of whey protein on the formation of pyrazines in Maillard model systems. *Food Chemistry*, *278*, 533–544. doi:10.1016/j.foodchem.2018.11.088 PMID:30583408

Schievano, E., Sbrizza, M., Zuccato, V., Piana, L., & Tessari, M. (2020). NMR carbohydrate profile in tracing acacia honey authenticity. *Food Chemistry*, *309*, 125788. doi:10.1016/j.foodchem.2019.125788 PMID:31753683

Sepahpour, S., Selamat, J., Khatib, A., Manap, M. Y. A., Abdull Razis, A. F., & Hajeb, P. (2018). Inhibitory effect of mixture herbs/spices on formation of heterocyclic amines and mutagenic activity of grilled beef. *Food Additives & Contaminants: Part A*, *35*(10), 1911–1927. doi:10.1080/19440049.2018.1488085 PMID:29913103

Serdaroğlu, M., Abdraimov, K., & Oenenc, A. (2007). The effects of marinating with citric acid solutions and grapefruit juice on cooking and eating quality of turkey breast. *Journal of Muscle Foods*, *18*(2), 162–172. doi:10.1111/j.1745-4573.2007.00074.x

Shamsudin, S., Selamat, J., Sanny, M., Jambari, N. N., Sukor, R., Praveena, S. M., & Khatib, A. (2020). The Inhibitory Effects of Heterotrigona Itama Honey Marinades on the Formation of Carcinogenic Heterocyclic Amines in Grilled Beef Satay. *Molecules (Basel, Switzerland)*, *25*(17), 3874. doi:10.3390/molecules25173874 PMID:32858787

Shapla, U. M., Solayman, M., Alam, N., Khalil, M., & Gan, S. H. (2018). 5-Hydroxymethylfurfural (HMF) levels in honey and other food products: Effects on bees and human health. *Chemistry Central Journal*, *12*(1), 1–18. doi:10.118613065-018-0408-3 PMID:29619623

Sharedeh, D., Gatellier, P., Astruc, T., & Daudin, J.-D. (2015). Effects of pH and NaCl levels in a beef marinade on physicochemical states of lipids and proteins and on tissue microstructure. *Meat Science*, *110*, 24–31. doi:10.1016/j.meatsci.2015.07.004 PMID:26172240

Sharif, M. K., Butt, M. S., Sharif, H. R., & Nasir, M. (2017). Sensory evaluation and consumer acceptability. Handbook of Food Science and Technology, 361–386.

Shehu, A., Ismail, S., Rohin, M. A. K., Harun, A., Abd Aziz, A., & Haque, M. (2016). Antifungal properties of Malaysian Tualang honey and stingless bee propolis against Candida albicans and Cryptococcus neoformans. *Journal of Applied Pharmaceutical Science*, *6*(2), 44–50. doi:10.7324/JAPS.2016.60206

Shin, H. S., & Ustunol, Z. (2004). Influence of honey-containing marinades on heterocyclic aromatic amine formation and overall mutagenicity in fried beef steak and chicken breast. *Journal of Food Science*, *69*(3), FCT147–FCT153. doi:10.1111/j.1365-2621.2004.tb13350.x

Singh, P., Yadav, S., Pathera, A., & Sharma, D. (2019). Effect of vacuum tumbling and red beetroot juice incorporation on quality characteristics of marinated chicken breast and leg meats. *Nutrition & Food Science*, *50*(1), 143–156. doi:10.1108/NFS-03-2019-0079

Smaoui, S., Ben Hlima, H., & Ghorbel, R. (2012). The effect of sodium lactate and lactic acid combinations on the microbial, sensory, and chemical attributes of marinated chicken thigh. *Poultry Science*, *91*(6), 1473–1481. doi:10.3382/ps.2011-01641 PMID:22582309

Sugimura, Y., Meyer, J., He, M. Y., Bart-Smith, H., Grenstedt, J., & Evans, A. G. (1997). On the mechanical performance of closed cell Al alloy foams. *Acta Materialia*, *45*(12), 5245–5259. doi:10.1016/S1359-6454(97)00148-1

Sulaiman, N. H. I., & Sarbon, N. M. (2022). Physicochemical, antioxidant and antimicrobial properties of selected Malaysian honey as treated at different temperature: A comparative study. *Journal of Apicultural Research*, *61*(4), 567–575. doi:10.1080/00218839.2020.1846295

Svečnjak, L., Bubalo, D., Baranović, G., & Novosel, H. (2015). Optimization of FTIR-ATR spectroscopy for botanical authentication of unifloral honey types and melissopalynological data prediction. *European Food Research and Technology*, *240*(6), 1101–1115. doi:10.100700217-015-2414-1

Szczesniak, A. S. (2002). Texture is a sensory property. *Food Quality and Preference*, *13*(4), 215–225. doi:10.1016/S0950-3293(01)00039-8

Szterk, A. (2015). Heterocyclic aromatic amines in grilled beef: The influence of free amino acids, nitrogenous bases, nucleosides, protein and glucose on HAAs content. *Journal of Food Composition and Analysis*, *40*, 39–46. doi:10.1016/j.jfca.2014.12.011

Tan, H. T., Rahman, R. A., Gan, S. H., Halim, A. S., Hassan, S. A., Sulaiman, S. A., & Bs, K.-K. (2009). The antibacterial properties of Malaysian tualang honey against wound and enteric microorganisms in comparison to manuka honey. *BMC Complementary and Alternative Medicine*, *9*(1), 1–8. doi:10.1186/1472-6882-9-34 PMID:19754926

Tänavots, A., Põldvere, A., Kerner, K., Veri, K., Kaart, T., & Torp, J. (2018). Effects of mustard-honey, apple vinegar, white wine vinegar and kefir acidic marinades on the properties of pork. *Veterinarija ir Zootechnika*, *76*(98), 76–84.

Tomac, A., Cova, M. C., Narvaiz, P., & Yeannes, M. I. (2015). Texture, color, lipid oxidation and sensory acceptability of gamma-irradiated marinated anchovy fillets. *Radiation Physics and Chemistry*, *106*, 337–342. doi:10.1016/j.radphyschem.2014.08.010

Tosi, E., Martinet, R., Ortega, M., Lucero, H., & Ré, E. (2008). Honey diastase activity modified by heating. *Food Chemistry*, *106*(3), 883–887. doi:10.1016/j.foodchem.2007.04.025

Traore, S., Aubry, L., Gatellier, P., Przybylski, W., Jaworska, D., Kajak-Siemaszko, K., & Sante-Lhoutellier, V. (2012). Higher drip loss is associated with protein oxidation. *Meat Science*, *90*(4), 917–924. doi:10.1016/j.meatsci.2011.11.033 PMID:22193037

Uckun, O., & Selli, S. (2017). Characterization of key aroma compounds in a representative aromatic extracts from citrus and astragalus honeys based on aroma extract dilution analyses. *Journal of Food Measurement and Characterization*, *11*(2), 512–522. doi:10.100711694-016-9418-9

Vallianou, N. G., Gounari, P., Skourtis, A., Panagos, J., & Kazazis, C. (2014). Honey and its anti-inflammatory, anti-bacterial and anti-oxidant properties. *General Medicine (Los Angeles, Calif.)*, *2*(132), 1–5. doi:10.4172/2327-5146.1000132

Van Haute, S., Raes, K., Van Der Meeren, P., & Sampers, I. (2016). The effect of cinnamon, oregano and thyme essential oils in marinade on the microbial shelf life of fish and meat products. *Food Control*, *68*, 30–39. doi:10.1016/j.foodcont.2016.03.025

Viegas, O., Moreira, P. S., & Ferreira, I. M. (2015). Influence of beer marinades on the reduction of carcinogenic heterocyclic aromatic amines in charcoal-grilled pork meat. *Food Additives & Contaminants: Part A*, *32*(3), 315–323. PMID:25604939

Viegas, O., Yebra-Pimentel, I., Martinez-Carballo, E., Simal-Gandara, J., & Ferreira, I. M. (2014). Effect of beer marinades on formation of polycyclic aromatic hydrocarbons in charcoal-grilled pork. *Journal of Agricultural and Food Chemistry*, *62*(12), 2638–2643. doi:10.1021/jf404966w PMID:24605876

Villalpando-Aguilar, J. L., Quej-Chi, V. H., López-Rosas, I., Cetzal-Ix, W., Aquino-Luna, V. Á., Alatorre-Cobos, F., & Martínez-Puc, J. F. (2022). Pollen Types Reveal Floral Diversity in Natural Honeys from Campeche, Mexico. *Diversity (Basel)*, *14*(9), 1–19. doi:10.3390/d14090740

Vlahova-Vangelova, D., & Dragoev, S. (2014). Marination: Effect on meat safety and human health. A review. *Bulgarian Journal of Agricultural Science*, *20*(3), 503–509.

Vlahova-Vangelova, D. B., Balev, D. K., Dragoev, S. G., & Kirisheva, G. D. (2016). Improvement of technological and sensory properties of meat by whey marinating. *Scientific Works of University of Food Technologies*, *63*(1), 7–13.

Wang, H., Wang, C., & Zhou, G. (2017). Effect of Polyphenols on Formation of Benzo (A) Pyrene in Charcoal-Grilled Meat. Jiangsu Synergetic Innovation Center of Meat Processing and Quality Control, 82(4), 684–690. doi:10.4315/0362-028X. JFP-18-420

Wang, H., Zhu, W., Ping, Y., Wang, C., Gao, N., Yin, X., Gu, C., Ding, D., Brinker, C. J., & Li, G. (2017). Controlled fabrication of functional capsules based on the synergistic interaction between polyphenols and MOFs under weak basic condition. *ACS Applied Materials & Interfaces*, *9*(16), 14258–14264. doi:10.1021/acsami.7b01788 PMID:28398036

Warner, R. D. (2023). The eating quality of meat: IV—Water holding capacity and juiciness. In *Lawrie's meat science* (pp. 457–508). Elsevier. doi:10.1016/B978-0-323-85408-5.00008-X

Wong, P., Ling, H. S., Chung, K. C., Yau, T. M. S., & Gindi, S. R. A. (2019). Chemical Analysis on the Honey of Heterotrigona itama and Tetrigona binghami from Sarawak, Malaysia. *Sains Malaysiana*, *48*(8), 1635–1642. doi:10.17576/jsm-2019-4808-09

World Health Organization. (2001). *Joint FAO/WHO Food Standard Programme Codex Alimentarius Commission Twenty-Fourth Session Geneva, 2-7 July 2001* (7th ed.). World Health Organization.

Woziwodzka, A., Gołuński, G., & Piosik, J. (2013). Heterocyclic aromatic amines heterocomplexation with biologically active aromatic compounds and its possible role in chemoprevention. *International Scholarly Research Notices*, *2013*, 1–11. doi:10.1155/2013/740821

Ya'akob, A., Norhisham, N. F., Mohamed, M., Sadek, N., & Endrini, S. (2019). Evaluation of physicochemical properties of trigona sp. stingless bee honey from various districts of Johor. *Jurnal Kejuruteraan*, *2*(1), 59–67.

Yanniotis, S., Skaltsi, S., & Karaburnioti, S. (2006). Effect of moisture content on the viscosity of honey at different temperatures. *Journal of Food Engineering*, *72*(4), 372–377. doi:10.1016/j.jfoodeng.2004.12.017

Yap, S. K., Chin, N. L., Yusof, Y. A., & Chong, K. Y. (2019). Quality characteristics of dehydrated raw Kelulut honey. *International Journal of Food Properties*, *22*(1), 556–571. doi:10.1080/10942912.2019.1590398

Yücel, B., Önenç, A., Bayraktar, H., Açikgöz, Z., & Altan, Ö. (2005). Effect of honey treatment on some quality characteristics of broiler breast meat. *Journal of Applied Animal Research*, *28*(1), 53–56. doi:10.1080/09712119.2005.9706788

Yusop, S. M., O'Sullivan, M. G., Kerry, J. F., & Kerry, J. P. (2010). Effect of marinating time and low pH on marinade performance and sensory acceptability of poultry meat. *Meat Science*, *85*(4), 657–663. doi:10.1016/j.meatsci.2010.03.020 PMID:20416811

Yusop, S. M., O'Sullivan, M. G., & Kerry, J. P. (2011). Marinating and enhancement of the nutritional content of processed meat products. In *Processed meats* (pp. 421–449). Elsevier. doi:10.1533/9780857092946.3.421

Zamora, R., & Hidalgo, F. J. (2020). Formation of heterocyclic aromatic amines with the structure of aminoimidazoazarenes in food products. *Food Chemistry*, *313*, 126128. doi:10.1016/j.foodchem.2019.126128 PMID:31951882

Zarei, M., Fazlara, A., & Tulabifard, N. (2019). Effect of thermal treatment on physicochemical and antioxidant properties of honey. *Heliyon*, *5*(6), e01894. doi:10.1016/j.heliyon.2019.e01894 PMID:31304409

Zeng, X., Ruan, D., & Koehl, L. (2008). Intelligent sensory evaluation: Concepts, implementations, and applications. *Mathematics and Computers in Simulation*, *77*(5–6), 443–452. doi:10.1016/j.matcom.2007.11.013

Zhang, L., Du, H., Zhang, P., Kong, B., & Liu, Q. (2020). Heterocyclic aromatic amine concentrations and quality characteristics of traditional smoked and roasted poultry products on the northern Chinese market. *Food and Chemical Toxicology*, *135*, 110931. doi:10.1016/j.fct.2019.110931 PMID:31678486

Zhang, Y., Song, Y., Zhou, T., Liao, X., Hu, X., & Li, Q. (2012). Kinetics of 5-hydroxymethylfurfural formation in chinese acacia honey during heat treatment. *Food Science and Biotechnology*, *21*(6), 1627–1632. doi:10.100710068-012-0216-9 PMID:31807335

Zhao, H., Cheng, N., Zhang, Y., Sun, Z., Zhou, W., Wang, Y., & Cao, W. (2018). The effects of different thermal treatments on amino acid contents and chemometric-based identification of overheated honey. *Lebensmittel-Wissenschaft + Technologie*, *96*, 133–139. doi:10.1016/j.lwt.2018.05.004

Zielbauer, B. I., Franz, J., Viezens, B., & Vilgis, T. A. (2016). Physical aspects of meat cooking: Time dependent thermal protein denaturation and water loss. *Food Biophysics*, *11*(1), 34–42. doi:10.100711483-015-9410-7

Zolkapli, M., Saharudin, S., Herman, S. H., & Abdullah, W. F. H. (2018). Quasi-distributed sol-gel coated fiber optic oxygen sensing probe. *Optical Fiber Technology*, *41*, 109–117. doi:10.1016/j.yofte.2017.12.016

Chapter 2

A Review on the Pollination Services by Stingless Bees, *Heterotrigona itama* (*Hymenoptera; Apidae; Meliponini*), on Some Important Crops in Malaysia

Wahizatul Afzan Azmi
Universiti Malaysia Terengganu, Malaysia

Wan Zaliha Wan Sembok
Universiti Malaysia Terengganu, Malaysia

Muhammad Firdaus Mohd. Hatta
Universiti Malaysia Terengganu, Malaysia

ABSTRACT

Stingless bees (Hymenoptera, Apidae, Meliponini) are common pollinators in the Malaysian agricultural ecosystem. Stingless bees are regarded as a good candidate for commercial pollination because of their specialized foraging adaptations and frequent visitation to cultivated fields. Unlike honeybees and bumble bees, stingless bees have not yet been commercially bred on a large scale for pollination purposes. Several studies outside Malaysia have shown that stingless bees' foraging activities may increase the production and quality of fruits. However, the role of stingless bees in producing quality fruits in open fields or in greenhouse crops in the Malaysian agricultural ecosystem is still unknown. In this review, the authors

DOI: 10.4018/978-1-6684-6265-2.ch002

discuss the efficiency of stingless bees, Heterotrigona itama, pollination services on some important cultivated crops in Malaysia, namely chili (Capsicum annuum), cucumbers (Cucumis sativus), and rock melon (Cucumis melo) based on previous reports. The findings revealed that pollination by H. itama can increase fruit size and weight, seed number, and pericarp volume.

INTRODUCTION

Chili *(Capsicum annuum)*, cucumber *(Cucumis sativus)* and rock melon *(Cucumis melo)* are some of the important cultivated crops in Malaysia and is a source of income for thousands of Malaysian farmers. According to Department of Statistics Malaysia (Monthly Manufacturing Statistics Malaysia, 2014), chili and cucumber average per capita consumptions (kg/year) in Malaysia were 2.0 kg/year and 2.9 kg/year in 2014, respectively. Compared to imported chili and cucumber, only small quantities of these vegetables were exported, and this shows that chili and cucumber productions in Malaysia still cannot meet the high demand by local consumer. Meanwhile, rock melon or cantaloupe is one of the commercialized fruits in Malaysia, where increasing popularity offers the potential for high profit returns to farmers. The Malaysian government has recognized that rock melon could generate significant economic income for farmers and has invested millions of Malaysia Ringgit in the industry (Griffith & Watson, 2016).

In Malaysia, chilies and cucumbers are usually grown outdoors, while rock melon is generally cultivated in greenhouses. However, due to better control of plant growth and environmental conditions, chilies and cucumbers are grown in the greenhouse where bees are excluded. Even though the flowers of these crops are self-pollinated, the anthers need to be shaken to allow effective pollen release. To overcome the problems, most farmers use many workers to help in manual cross-pollination of the flowers which causes increase in hiring costs and time needed to pollinate the flowers. Thus, it has been suggested to use the stingless bees as the pollination agent in the greenhouse-cultivated crops as there are many researchers found that pollination by stingless bees can increase fruit size and weight, pericarp volume, high percentage of seed per fruits and fastens harvesting time (Slaa et al., 2006; Nicodemo et al., 2013; Nunes-Silva et al., 2013).

Stingless bee is one of the important pollinators in open field crops as well as in greenhouse crops (Heard, 1999). Even though stingless bees are common pollinators in Malaysian agricultural ecosystem, their role in pollination success is still poorly understood. Certain species of stingless bees are regarded as candidates for commercial pollination because of their specialized foraging adaptations and frequent visitation to cultivated fields (Slaa et al., 2006). Several studies on strawberry (*Fragaria ananassa*), tomato (*Lycopersicon esculentum*) (Del Sarto et al., 2005; Nunes-Silva et al., 2013), eggplant (*Solanum melongena*), sweet pepper (*Capsicum annum*) (Cruz et al., 2005) and cucumber (*Cucumis sativus*) (dos Santos et al., 2008). Nicodemo et al. (2013) had shown that stingless bees foraging activities may increase the production as well as the quality of fruits.

Even though considerable studies have been conducted on the use of stingless bees for commercial pollination purposes, however to date, very little attention has been made to study the pollination efficiency of the stingless bees specifically in Malaysia region. The efficiency of native stingless bees particularly *Heterotrigona itama*, however, is poorly studied and need to be proven. It has been reported that the stingless bees could enhance seed production by raising the amount of mature pollen grains by transferring them to other plants' stigmas (Cauich et al., 2006). Therefore, this study is aimed to highlight the pollination efficiency by our native stingless bee, *H. itama* on the quality and production of chili, cucumber and rock melon grown in the greenhouses. This review paper is basically a compilation of the findings from previous studies by Azmi et al. (2016; 2017; 2019).

MATERIALS AND METHODS

The Materials and Methods described in this paper are based on Azmi et al. (2016), (2017), (2019). Below are the summarized methods from the studies:

Plant Materials and Experimental Location

The study was carried out in 6.0 x 24.0 m^2 greenhouses at the Faculty of Fishery and Food Science, Universiti Malaysia Terengganu (N05^{0}24.541', E103^{0}05.347'). Chili, cucumber and rock melon seeds were provided by the Department of Agriculture of Terengganu State. Seeds were initially soaked in a mixed solution of water and Thiram (seed treatment) for 15 to 25 minutes to prevent fungal

43

diseases in seeds. Then, the seeds were rinsed for about 5 minutes under running water and were sown in shallow seed trays with moist compost. In about 3 to 4 weeks, the seedlings were transplanted into the polybags (60 x 40 cm) containing 5 kg of coco peat. Plants were irrigated daily by drip irrigation.

Rearing of Stingless Bees, *Heterotrigona itama*

Three strong colonies of wild stingless bees, *H. itama* were obtained from the local suppliers in Terengganu. The stingless bees were transferred from their natural habitat into the man-made hives. The hives were built from wooden box of 60 cm x 25 cm x 10 cm. The stingless bee colonies were maintained and acclimatized for approximately 1 month in the field under the natural climatic conditions ($\pm$ 22 to 31°C, 83% humidity).

Experimental Design

The greenhouses were divided into three equal parts by using approximately 2 m of mist net with the opening size of 250 x 720 μm, where one third was occupied by plots using self-pollination and the other two parts were occupied by the remaining treatments which were hand-cross pollination and stingless bee polllination, respectively. The treatments were (1) self-pollination as control, (2) hand-cross pollination and (3) stingless bees (*H. itama*) pollination with three replications each. In the self-pollination (control), 20 flower buds (for each crop species) were chosen randomly and were bagged with fine muslin bags (mesh 1 x 1 mm) until fruit set to prevent the flowers from being pollinated by other pollination agents. The flowers were individually tagged. In the hand-cross pollination, 20 flower buds were selected, tagged and bagged. After the anthesis of flowers at 0700 to 1000 h, the flowers were unbagged and hand-pollinated with pollen taken from flowers of other plants by gently rubbing the flower stigma with a stamen containing fresh dehisced pollen grains. Then the flowers were rebagged to avoid other pollinators and properly tagged. In the pollination treatment by stingless bees, three hives of strong colonies of *H. itama* were placed at the centre of the greenhouse for at least two days before the anthesis period in order for the stingless bees to adapt with the new environment. Twenty flowers that received one *H. itama* visit were tagged and bagged until fruit set. Figure 1 shows the photos of pollination treatment by *H. itama* on chili, cucumber and rock melon flowers.

Figure 1. Pollination by stingless bee, Heterotrigona itama on flowers of chili

Figure 2. Cucumber

Figure 3. Rock melon

Production-Related Parameters

After about 4 to 5 weeks, the matured fruits were harvested. Six parameters were evaluated in this experiment: fruit weight, fruit size (diameter and length), average number of seeds per fruit, fruit firmness and Total Soluble Solid (TSS). The fruit weight (g) was measured using an electronic weight balance, the diameter (mm) and length (cm) measured using vernier caliper. The fruit firmness was measured using a Texture Analyzer (Model TA1, Lloyd, UK) with probe 2N. The Total Soluble Solid (TSS) content was measured using a hand-held Refractometer (Model MT-032, Reed Instruments, UK).

Statistical Analysis

All the data obtained from this study were analyzed using statistical software produced by IBM Statistical Package for the Social Sciences (SPSS) version 21.0. One-way analysis of variance (ANOVA) was used to test whether or not there are significant differences in each parameter measured among the treatments. When there were significant differences, the Tukey post-hoc test was applied to determine which means were most alike (or different) and to test the equality of means for each pair of variables.

RESULTS AND DISCUSSION

Pollination Efficiency of *H. itama* on Chilies

Results of production-related parameters clearly showed that chilies grown in the greenhouse benefit from pollination by *H. itama*. Azmi et al. (2016) showed that chilies produced from pollination by the stingless bees, *H. itama* and hand-cross pollination were significantly heavier, longer and containing greater number of seeds per fruit than self-pollinated chilies (p<0.05) (see Table 1). Cruz et al. (2005) reported that higher number of seed developing inside sweet pepper fruits produced from stingless bees, *Melipona subnitida* pollination would lead to bigger and heavier fruits in the greenhouse. Seeds play an important role in fruit setting process, since poorly developed fruits are the result of an unequal seed distribution inside the fruit (McGregor, 1976). Similar result was obtained by Shipp et al. (1994) where they obtained heavier and bigger sweet pepper fruits produced from the bumble bee pollination. Therefore, it is clearly shown that the quality of chilies from our native stingless bee pollination, *H. itama* was almost similar to the chilies produced from hand-cross pollination, but better than self-pollination treatment.

Table 1. Comparison of production-related parameters between chilies produced from self-pollination, hand-cross pollination and H. itama pollination

Treatment	No. of Seeds (no/ fruit)	Weight (g)	Length (cm)	Diameter (mm)	Total Soluble Solid (°)	Firmness (N)
Self-pollination	48.54 ± 15.28[a]	8.63 ± 1.45[a]	9.02 ± 0.77[a]	1.38 ± 1.28[a]	9.64 ± 1.07[a]	1.702 ± 0.30[a]
Hand-cross pollination	102.92 ± 24.25[b]	9.77 ± 1.54[ab]	9.23 ± 1.89[a]	11.58 ± 1.05[a]	9.48 ± 2.18[a]	1.789 ± 0.31[a]
H. itama pollination	112.54 ± 21.15[b]	1.61 ± 0.86[b]	13.00 ± 0.59[b]	12.40 ± 0.94[a]	7.35 ± 0.51[a]	1.839 ± 0.37[a]

Note: *Means + SD in column with same letter are not significantly different (p>0.05) according to Tukey Post Hoc test (n = 20).

Source: Azmi et al. (2016).

Pollination Efficiency of *H. itama* on Cucumbers

Results in the study by Azmi et al. (2017) showed that the cucumbers pollinated by stingless bee and hand-cross pollination produced heavier, longer and larger cucumbers compared to those produced from pollination without stingless bees (p<0.05). However, in terms of dried weight seed and firmness of cucumbers, no significant difference between the treatments were detected (p>0.05) (see Table 2). Similar results were obtained with previous study by Nicodemo et al. (2013), where they found that the pollination by Brazilian native stingless bee contributed to a significant increase in cucumber diameter and length compared to the cucumbers pollinated by honeybees. In addition, cucumbers are monoecious plants which have separate female and male flowers in the same plant. Thus, it needs the bees as pollination agent. As stated by Nicodemo et al. (2013), the production of cucumber increased by 26% when the stingless bees were placed in the greenhouse. This shows that cucumber flowers pollinated by the pollination agents such as stingless bees have better qualities in terms of weight, length and diameter.

Table 2. Comparison of production-related parameters between cucumbers produced from self-pollination, hand-cross pollination and H. itama pollination

Treatment	Dried Seed Weight (g)	Length (cm)	Weight	Diameter	Firmness (N)
			(g)	(cm)	
Self-pollination	0.46 ± 0.23[a]	19.61 ± 1.62[a]	0.30 ± 0.09[a]	15.75 ± 2.07[a]	26.49 ± 4.51[a]
Hand-cross pollination	0.82 ± 0.32[b]	17.58 ± 1.21[a]	0.42 ± 0.07[ab]	17.58 ± 1.21[b]	25.42 ± 3.16[b]
H. itama pollination	0.79 ± 0.25[b]	22.20 ± 1.12[b]	0.43 ± 0.08[b]	17.84 ± 1.01[b]	23.85 ± 2.12[a]

Note: *Means + SD in column with same letter are not significantly different (p>0.05) according to Tukey Post Hoc test (n = 20).

Source: Azmi et al. (2017)

Pollination Efficiency of *H. itama* on Rock Melons

Findings from Azmi et al. (2019) revealed that rock melons produced from pollination by *H. itama* and hand-cross pollination treatments were similar in number of fruits sets, seeds, weight and diameter (Table 3). Roselino et al. (2009) who used the stingless bee species *Scaptotrigona aff. depilis* and *Nannotrigona testaceicornis* as pollinating agents and found that quality of fruit produced from flowers pollinated by the stingless bees was higher than in fruit produced from self-pollination.

Similarly, dos Santos et al. (2008) reported that crops pollinated by the stingless bee *Melipona quadrifasciata* produced higher quality, heavier, and larger fruit than crops pollinated without the species. It has been suggested that stingless bee behavior could affect the weight and diameter of fruits, since they spend more time collecting pollen and the nectar (Roselino et al., 2009). Interestingly, rock melons pollinated by *H. itama* produced significantly greater sweetness than self-pollination and hand-cross pollination ($p<0.05$). The Total Soluble Solid (TSS) of rock melon in the stingless bee pollination treatment was higher in the hand cross-pollination and self-pollination treatments (see Table 3). Pollination by bees has been shown to increase auxin and gibberellic acid production (Slaa et al., 2006). Thus, it shows that stingless bee pollination improves the quality of rock melon and the sweetness of the fruit.

Table 3. Comparison of production-related parameters between rock melons produced from self-pollination, hand-cross pollination and H. itama pollination

Treatment	No. of Fruit Sets (no/ plant)	No. of Seeds (no/ fruit)	Weight (g)	Diameter (mm)	Total Soluble Solid (°)	Firmness (N)
Self-pollination	13.00 ± 7.55[a]	494.00 ± 53.31[a]	1.35 ± 0.21[a]	13.57 ± 1.05[a]	8.64 ± 1.39[a]	13.48 ± 2.06[a]
Hand-cross pollination	22.67 ± 10.60[ab]	554.56 ± 49.53[ab]	1.53 ± 0.17[ab]	14.63 ± 0.36[b]	9.95 ± 1.16[a]	13.42 ± 1.82[a]
H. itama pollination	37.33 ± 6.03[b]	613.89 ± 72.47[b]	1.71 ± 0.22[b]	15.32 ± 0.43[b]	11.01 ± 1.35[b]	12.02 ± 0.76[a]

Note: *Means ± SD in column with same letter are not significantly different ($p>0.05$) according to Tukey Post Hoc test (n = 20).
Source: Azmi et al. (2019)

CONCLUSION

As a conclusion, chilies, cucumbers and rock melons which resulted from pollination by stingless bee *H. itama* have improved in terms of crop quantity and fruit quality where heavier, longer, higher seed number and percentage of fruit sets were produced. In contrast, hand-cross pollination needs a lot of precision and requires skilled workers in order to produce high quality crops. Failure to do so may probably lead to high number of malformed fruits. Thus, this study concluded that stingless bee, *H. itama* contributed to higher fruit quality compared to that of the pollination without stingless bee. This reveals the potential of *H. itama* to be utilized as an alternative pollinator in agricultural ecosystem in Malaysia.

ACKNOWLEDGMENT

Authors would like to acknowledge Unit Peneraju Agenda Bumiputera (TERAJU) for the financial support under Dana Pembangunan Usahawan Bumiputera (DPUB) Terengganu (Vot: 58920). Also, the authors are indebted to the Universiti Malaysia Terengganu for providing the laboratory facilities.

REFERENCES

Azmi, W. A., Samsuri, N., Hatta, M. F. M., Ghazi, R., & Seng, C. T. (2017). Effects of stingless bee (Heterotrigona itama) pollination on greenhouse cucumber (Cucumis sativus). *Malaysian Applied Biology*, *46*(1), 51–55.

Azmi, W. A., Seng, C. T., & Solihin, N. S. (2016). Pollination efficiency of the stingless bee, Heterotrigona itama (Hymenoptera: Apidae) on chili (Capsicum annuum) in greenhouse. *J. Trop. Plant Physiol*, *8*, 1–11.

Azmi, W. A., Wan Sembok, W. Z., Yusuf, N., Mohd. Hatta, M. F., Salleh, A. F., Hamzah, M. A. H., & Ramli, S. N. (2019). Effects of pollination by the indo-Malaya stingless bee (Hymenoptera: Apidae) on the quality of greenhouse-produced rockmelon. *Journal of Economic Entomology*, *112*(1), 20–24. doi:10.1093/jee/toy290 PMID:30277528

Cauich, O., Quezada Euan, J. J. G., Ramírez, V. M., Valdovinos-Nuñez, G. R., & Moo-Valle, H. (2006). Pollination of habanero pepper (*Capsicum chinense*) and production in enclosures using the stingless bee *Nannotrigona perilampoides*. *Journal of Apicultural Research*, *45*(3), 125–130. doi:10.1080/00218839.200 6.11101330

Cruz, D. de O., Freitas, B. M., da Silva, L. A., da Silva, E. M. S., & Bomfim, I. G. A. (2005). Pollination efficiency of the stingless bee Melipona subnitida on greenhouse sweet pepper. *Pesquisa Agropecuária Brasileira, 40*(12), 1197–1201. doi:10.1590/S0100-204X2005001200006

Del Sarto, M. C. L., Peruquetti, R. C., & Campos, L. A. O. (2005). Evaluation of the neotropical stingless bee Melipona quadrifasciata (Hymenoptera: Apidae) as pollinator of greenhouse tomatoes. *Journal of Economic Entomology, 98*(2), 260–266. doi:10.1093/jee/98.2.260 PMID:15889711

dos Santos, S. A. B., Roselino, A. C., & Bego, L. R. (2008). Pollination of cucumber, Cucumis sativus L.(Cucurbitales: Cucurbitaceae), by the stingless bees Scaptotrigona aff. depilis moure and Nannotrigona testaceicornis Lepeletier (Hymenoptera: Meliponini) in greenhouses. *Neotropical Entomology, 37*(5), 506–512. doi:10.1590/S1519-566X2008000500002 PMID:19061034

Griffith, G., & Watson, A. (2016). Agricultural markets and marketing policies. *The Australian Journal of Agricultural and Resource Economics, 60*(4), 594–609. doi:10.1111/1467-8489.12161

Hasnol, N. D. S., Jinap, S., & Sanny, M. (2014). Effect of different types of sugars in a marinating formulation on the formation of heterocyclic amines in grilled chicken. *Food Chemistry, 145*, 514–521. doi:10.1016/j.foodchem.2013.08.086 PMID:24128508

Heard, T. A. (1999). The role of stingless bees in crop pollination. *Annual Review of Entomology, 44*(1), 183–206. doi:10.1146/annurev.ento.44.1.183 PMID:15012371

McGregor, S. E. (1976). Insect pollination of cultivated crop plants (Issue 496). Agricultural Research Service, US Department of Agriculture.

Monthly Manufacturing Statistics Malaysia. (2014). Department of Statistics.

Nicodemo, D., Malheiros, E. B., De Jong, D., & Nogueira Couto, R. H. (2013). Enhanced production of parthenocarpic cucumbers pollinated with stingless bees and Africanized honey bees in greenhouses. *Semina. Ciências Agrárias, 34*(6Supl1), 3625–3633. doi:10.5433/1679-0359.2013v34n6Supl1p3625

Nunes-Silva, P., Hrncir, M., Shipp, L., Kevan, P., & Imperatriz-Fonseca, V. L. (2013). The behaviour of Bombus impatiens (Apidae, Bombini) on tomato (Lycopersicon esculentum Mill., Solanaceae) flowers: Pollination and reward perception. *Journal of Pollination Ecology, 11*, 33–40. doi:10.26786/1920-7603(2013)3

Roselino, A. C., Santos, S. B., Hrncir, M., & Bego, L. R. (2009). Differences between the quality of strawberries (Fragaria x ananassa) pollinated by the stingless bees Scaptotrigona aff. depilis and Nannotrigona testaceicornis. *Genetics and Molecular Research, 8*(2), 539–545. doi:10.4238/vol8-2kerr005 PMID:19551642

Shipp, J. L., Whitfield, G. H., & Papadopoulos, A. P. (1994). Effectiveness of the bumble bee, Bombus impatiens Cr.(Hymenoptera: Apidae), as a pollinator of greenhouse sweet pepper. *Scientia Horticulturae, 57*(1–2), 29–39. doi:10.1016/0304-4238(94)90032-9

Slaa, E. J., Chaves, L. A. S., Malagodi-Braga, K. S., & Hofstede, F. E. (2006). Stingless bees in applied pollination: Practice and perspectives. *Apidologie, 37*(2), 293–315. doi:10.1051/apido:2006022

Yusop, S. M., O'Sullivan, M. G., Kerry, J. F., & Kerry, J. P. (2010). Effect of marinating time and low pH on marinade performance and sensory acceptability of poultry meat. *Meat Science, 85*(4), 657–663. doi:10.1016/j.meatsci.2010.03.020 PMID:20416811

Chapter 3

Antimicrobial Activity From Five Species of Stingless Bee (*Apidae meliponini*) Honey From South East Asia (Thailand)

Jakkrawut Maitip
King Mongkut's University of Technology North Bangkok, Thailand

Sirikarn Sanpa
University of Phayao, Thailand

Michael Burgett
Oregon State University, USA

Bajaree Chuttong
Chiang Mai University, Thailand

ABSTRACT

In Thailand, there have been limited investigations on the antibacterial properties of stingless bee honey. The purpose of this research is to investigate the physicochemical and antibacterial characteristics of five stingless bee species, including Lepidotrigona flavibasis, L. doipaensis, Lisotrigona furva, Tetragonula laeviceps species complex, and T. testaceitarsis complex from two geographical locations in Thailand: North (Chiang Mai) and Southeast (Chanthaburi). The moisture content from five species of stingless bee ranged from 27.6 to 32.0 g/100g. The range of pH in stingless bee

DOI: 10.4018/978-1-6684-6265-2.ch003

honey was 3.5 to 3.8, which is slightly lower than the pH of Apis mellifera honey. The total acidity of stingless bee honey ranged from 44.0 to 216.9 meq/kg. The antimicrobial property of honey samples was investigated by the agar disc-diffusion method followed by MIC/MBC assay. Notably, with the exception of L. furva, stingless bee honeys were shown to exhibit antibacterial against the Gram-negative bacteria greater than Gram-positive bacteria.

INTRODUCTION

Most tropical and subtropical parts of the world possess stingless bees. There are now over 500 species described (Michener, 2013). The Neotropics of Central and South America have the richest diversity of meliponines, with 400 recognized species. The remaining species are found in Africa and Indo-Australian regions (Cortopassi-Laurino et al., 2006; Michener, 2013; Pauly et al., 2013). For Thailand, (Rasmussen, 2008) listed 32 species within 10 genera. Humans have been keeping stingless bees for centuries with some species being managed before A. mellifera was distributed worldwide (Cortopassi-Laurino et al., 2006). Stingless bee beekeeping is known as meliponiculture. In Thailand, it is considered to be in a developmental state (Chuttong et al., 2014). Stingless bees have been known for their honey, pollen and wax (cerumen) production. Comparing honey production between the stingless bees and A. mellifera (the western honey bee), honey is insignificant. Stingless bee honey is not yet included in international standards and establishing such a standard requires a better understanding of its composition and qualities (Chuttong et al., 2016; Souza et al., 2006). Tetragonula laeviceps species complex is the most predominant species reared in a box hives in Thailand. T. testaceitarsis, T. fuscobalteata, L. flavibasis, L. doipaensis, and L. furva are among the other species managed in wooden boxes or horizontal and vertical log hives.

Honey contains sugars and other constituents such as enzymes, amino acids, organic acids, carotenoids, vitamins, minerals, and aromatic substances. The composition and quality of honey depend on its nectar origin, honey maturity, production methods, climatic conditions, and processing and storage conditions (Bogdanov et al., 2008; Nombré et al., 2010; White & Doner, 1980). The properties of different types of honey have been indicated by many scientists (Bogdanov et al., 2008). Most of the previous research on honey quality and properties has concentrated on the western honey bee (A. mellifera). According to the European Codex Honey Standards by Alimentarius (2001), the primary criteria to determine honey's quality are moisture content, ash content, electrical conductivity, carbohydrates, acidity, diastase activity, and hydroxymethylfurfural content.

Honey has traditionally been used to treat wounds and inflammations. The antibacterial action of honey is ascribed to its low pH and high osmolality, which are sufficient to inhibit microbial proliferation. Much research has been conducted to evaluate the antibacterial activity against pathogenic bacteria, oral bacteria, and food spoilage bacteria of honey (Alvarez-Suarez et al., 2010; Escuredo et al., 2012; Kwakman et al., 2010; Lusby et al., 2005; Vallianou et al., 2014; White & Doner, 1980). Honey is also known for the enzymatic production of hydrogen peroxide and other non-peroxide factors as lysozymes, phenolic acids and flavonoids. Honey exhibits a wide range of biological effects and natural antioxidants (Alvarez-Suarez et al., 2010).

Stingless bee honey has a long consumption history due to its composition, nutritional characteristics, and beneficial medicinal uses, particularly in South America, Australia, Asia and Africa (Guerrini et al., 2009; Vit et al., 1994). There are several studies on honey's antibacterial properties in vitro. Ethiopian stingless bee honey prevented the growth of wound-infecting bacteria such as Escherichia coli and Staphylococcus aureus. Melipona compressipes manaosensis honey from Brazil demonstrated the effectiveness against E. coli, Proteus vulgaris and Klebsiella sp. (McLoone et al., 2016). Antibacterial activity of stingless bee honey from Costa Rica indicated activity against several bacterial strains (Zamora et al., 2013). Only one species of stingless bee honey (Tetragonula laeviceps) has been studied for antibacterial activity in Thailand (Chanchao, 2009). Chanchao (2013) investigated the properties of T. laeviceps honey, exhibiting inhibitory zones against two bacterial species (S. aureus and E. coli), two species of yeast (Candida albicans, Auriobasidium pullalans) and the fungi Aspergillus niger. (Suntiparapop et al., 2012) investigated the antibacterial activity of T. laeviceps honey purchased from meliponine apiaries and discovered that the honey inhibited the development of various bacteria except for Propionibacterium acnes and two species of yeast. (Saccharomyces cerevisiae and C. albicans). The purpose of this study was to examine the antimicrobial effectiveness of stingless bee honey from five species obtained from natural and human-made hives in the Thai provinces of Chanthaburi and Chiang Mai against nine species of pathogenic bacteria. Previous research has reported the antimicrobial attributes from only one species of stingless bee honey from Thailand.

MATERIALS AND METHODS

Sample Collection

Stingless bee honey was collected from five species of stingless bees. Honey samples from each species were obtained from three different colonies. Samples were taken from sealed honey pots, pierced with a sharp tool and the honey was strained through a fine cloth. Honey samples were collected from two geographical locations in Thailand: North (Chiang Mai) 2 species of stingless bee (L. flavibasis and L. doipaensis) and Southeast (Chanthaburi) 3 species of stingless bee (T. testaceitarsis complex, T. laeviceps species complex and L. furva). Adult worker stingless bee specimens were collected for taxonomic identification. The A. mellifera sample was taken from an apiary located in Lamphun province, with the primary nectar source being longan (Dimocarpus longan). Honey samples were stored at 4°C in sealed glass jars in the dark.

Moisture

Honey moisture was determined by refractometry (AOAC method 919.38, 2006) using an Atago (Japan) model N-3E refractometer. All measurements were performed at 20 °C in triplicate.

pH and Total Acidity

Honey pH and total acidity were determined according to AOAC method 962.19 (2006). Ten grams of honey was dissolved in 75 ml distilled water then the solution was titrated with 0.05M NaOH solution until a pH of 8.5. Ten ml of 0.05M NaOH was added immediately, then back titrated with 0.05M HCl solution until the pH reached 8.3 to determine the acidity. A Cyberscan waterproof (Singapore) model PC300 Series digital pH meter was used. The analyses were performed in triplicate.

Antimicrobial Activity Test

The antibacterial activity of honey samples was investigated using nine bacterial strains, five of which were gram-negative and four of which were gram-positive. The microorganisms obtained from the Thailand Institute of Scientific and Technological Research, were Enterobacter aerogenes (TISTR 1540), Escherichia coli (TISTR 527), Pseudomonas aeruginosa (TISTR 1287), Salmonella choleraesuis (TISTR 1481), Salmonella typhimurium (TISTR 1469), Bacillus subtilis (TISTR 008), Clostridium butyricum (TISTR 1032), Staphylococcus aureus (TISTR 1840) and Staphylococcus epidermis (TISTR 518). In order to reach the stationary growth phase, C. butyricum was cultivated in cooked meat medium (R.C. medium) under anaerobic conditions, S. epidermis in tryptic soy broth (TSB), and other bacteria in nutrient broth (NB) under aerobic conditions. All bacteria were cultured at 37 °C for 16-18 h.

Ager Disc-diffusion Method

Antimicrobial tests were carried out by the agar disc-diffusion method (Dajanta et al., 2012). Agar plates were inoculated with 200 µl of the test microorganism (106 CFU/ml) suspensions. Then, filter paper disks (6 mm in diameter; paper chromatography, Whatman No.1) were saturated with 20 µl of the honey samples and placed onto the inoculated plates. The Petri dishes were incubated at 37 °C for 16-18 h. A digital vernier caliper measured the diameters (mm) of the inhibition growth zones. A. mellifera honey was used as positive control and the negative control was sterilized distilled water.

Minimum Inhibitory Concentration (MIC) and Minimum Bactericidal Concentration (MBC) Assay

The MIC was determined using the broth dilution method. All microorganisms used to determine minimum inhibitory concentrations (MICs) were freshly prepared before testing. Each culture was transferred to a sterile phosphate-buffered saline (PBS), and turbidity was adjusted by 0.5 McFarland Standard following Clinical and Laboratory Standards Institute (CLSI) guidelines (Wikler, 2006) with slight modifications (Maitip et al., 2021). The inhibition of microbial growth was determined by measuring the absorbance at 600 nm. The MIC was defined as the lowest concentration of stingless bee honey that completely inhibited the development of microorganisms and was indicated by the absence of turbidity in the broth. The MBC was obtained using the lowest dose of stingless bee honey, which killed 99.9% of the test microorganisms when sub-cultured on agar plates. The analyses were carried out in triplicate. The MIC and MBC values were determined using stingless bee honey samples from four species that demonstrated the inhibitory zone of bacteria using the agar-disc diffusion technique.

Statistical Analyses

The data were reported as mean ± standard deviations (SDs) from three replicates in each experiment. The data were compared with One-way ANOVA with Tukey's pairwise comparisons. If the normality assumption was violated, Kruskal–Wallis test with Mann–Whitney pairwise comparisons were applied. Data with P values smaller than 0.05 were considered statistically significant. All statistical analyses were performed using NCSS software version 2021 (Cho & Tony Ng, 2021).

RESULTS AND DISCUSSION

The examination of stingless bee honey moisture, pH and acidity from five species are presented in Table 1.

Table 1. Moisture, pH and acidity of five species of stingless bees compared to Apis mellifera honey

Species	Location	Moisture (g/100g)	pH	Acidity (meq/kg)
Lepidotrigona flavibasis	Chiang Mai	$30.3^b \pm 1.5$	3.6 ± 0.1	$194.5^a \pm 24.9$
Lepidotrigona doipaensis	Chiang Mai	$31.7^a \pm 0.7$	3.5 ± 0.1	$198.3^a \pm 9.6$
Lisotrigona furva	Chanthaburi	$28.2^c \pm 0.6$	3.6 ± 0.1	$53.1^b \pm 5.2$
Tetragonula laeviceps species complex	Chanthaburi	$28.2^c \pm 0.8$	3.6 ± 0.2	$49.5^c \pm 5.2$
Tetragonula testaceitarsis complex	Chanthaburi	$30.7^b \pm 1.3$	3.6 ± 0.1	$69.3^b \pm 16.5$
Apis mellifera	Lamphun	21.7^d	3.9	18.5^d

Note: Column values that do not share a common letter are significantly different at p £ 0.05. (One-Way ANOVA followed by post hoc Turkey test). Columns without letters display no significant difference.

Moisture

Moisture content from five species of stingless bee ranged from 27.6 to 32.0 g/100g. L. furva and T. laeviceps showed the lowest moisture content (27.7 to 28.9 and 27.6 to 29.1g/100g) while L. doipaensis displayed the highest moisture content (31.1 to 32.4 g/100g). Compared to previous studies of stingless bee honey moisture content from comparable species, our research indicates similar results (Chuttong et al., 2016; Suntiparapop et al., 2012). The moisture content of thirty-one species of stingless bee honey from South America described by Patricia Vit et al. (2013) revealed the moisture content range from 16.5 to 42.7 g/100g. Stingless bee honey frequently possesses a higher moisture content than the western honey bee (A. mellifera) (Souza et al., 2006), as our result show in Table 1. Honey is a saturated sugar solution, with the moisture content of honey (A. mellifera) usually being 15-21%. The property of high sugars and low water content of honey, called the osmotic effect, allows very limited water available for microorganisms. However, the results from some experiments have explained that an osmotic effect is not the only factor involved in the antibacterial activity of honey (Molan, 1992).

pH and Total Acidity

Our study examined the pH of fifteen samples of stingless bees and a sample of A. mellifera honey. The range of pH found in stingless bee honey was 3.5 to 3.8. It was slightly lower than the pH of A. mellifera honey. The total acidity of stingless bee honey ranged from 44.0 to 216.9 meq/kg, which demonstrates the variation between species. L. furva, T. laeviceps complex and T. testaceitarsis complex demonstrated lower acidity than the other two species of stingless bee honey but still higher than A. mellifera honey. When comparing the pH and acidity of the individual species of stingless bee honey to the previous reports, Chuttong et al. (2016) and Suntiparapop et al. (2012) stated that pH from the same species of stingless bee honey falls in the same range as our results reported here. However, Chanchao (2009) investigating the pH of T. laeviceps reported a honey pH of 3.37, which is lower. Then again, the total acidity content found in our honey samples was related to (Chuttong et al., 2016). However, the report of Suntiparapop et al. (2012) for T. laeviceps honey showed, the mean total acidity was 72.84 meq/kg. The acidity value relates to the organic acids exhibited in honey and can vary according to the floral origin and bee species (Biluca et al., 2016). Bogdanov (1997) revealed the correlation between acidity, pH and antibacterial activity of honey. The results showed the bacterial inhibition was significantly correlated to the free and total acidity but no correlation with the pH of honey. From his study it can be assumed that the acidity in honey exerts a primary antibacterial action, while honey pH could perform as an antibacterial factor.

ANTIBACTERIAL ACTIVITY ANALYSIS

Agar Disc Diffusion Method

The antibacterial properties of five species of stingless bee honey compared to A. mellifera honey, were examined by the agar disc diffusion method. The inhibition zone diameters of honey samples against gram-positive and gram-negative bacterial species are presented in Table 2. The results reveal that the antibacterial property of stingless be honey samples displayed broad-spectrum activity but were more effective against Gram-negative compared to Gram-positive bacteria. The inhibition of gram-negative bacteria was revealed in four species of stingless bee honey (L. flavibasis, L. doipaensis, T. testaceitarsis, and T. laeviceps species complex), the inhibition zone against five strains of Gram-negative bacteria ranged from 6.5 – 14.0

mm except for L. furva honey. In comparison, the inhibition zone of honey against four strains of Gram-negative bacteria was ranged from 6.9 – 22.5 mm. However, each honey bee sample was only able to inhibit three out of four Gram-negative bacteria strains.

Overall, based on the statistical analysis, honey most active against bacteria in this study belonged T. testaceitarsis, and T. laeviceps species complex, which inhibited eight out of ten strains of bacteria, followed by L. flavibasis, and L. doipaensis inhibiting seven out of ten strains of bacteria. Whereas, the honey from L. furva was not able to inhibit any tested bacteria. For Gram-negative bacteria inhibition, honey from L. flavibasis showed the greatest inhibition (inhibition zone) against E .aerogenes, E. coli, P. aeruginosa, S. choleraesuis, and S. typhimuruim (12.2±1.3, 14.0±1.0, 13.7±2.4, 10.3±1.5, and 13.0±2.1 mm, respectively) followed by L. doipaensis (13.2±2.8, 10.3±1.0, 12.2±1.0, 7.8±0.8, and 8.2±0.6, respectively), T. laeviceps complex (10.7±1.6, 9.5±2.3, 8.1±0.5, 8.5±1.0, and 8.7±1.2, respectively), and T. testaceitarsis (7.8±1.1, 6.5±0.0, 7.5±0.0, 8.0±0.0, and 10.7±0.2, respectively). In contrast, the antibacterial activity of honey samples against Gram-positive bacteria were lower than Gram-negative bacteria. C. botulinum was susceptible to all honey samples excepted L. furva and also showed the greatest inhibition (inhibition zone) (22.5±0.0, 17.0±2.0, 10.5±0.5, and 10.2±0.8: L. doipaensis, L. flavibasis, T. testaceitarsis, and T. laeviceps complex, respectively). S. epidermis was susceptible to honey from T. laeviceps complex and T. testaceitarsis (18.3±3.2 and 13.0±1.0, respectively). S. aureus was susceptible only to honey from T. testaceitarsis and L. doipaensis (12.0±1.0 and 6.9±0.2, respectively) and B. subtilis was susceptible to honey from L. flavibasis (6.9±0.2) and T. laeviceps complex (7.7±0.3).

Chanchao (2009) examining the inhibition of T. laeviceps honey to E. col,i found the clear zone diameter was 29.4 mm (the clear zone diameter measurement was performed after 72 h of culture), which is more significant than our results. With the report of Chanchao (2013), the inhibition zone diameter of T. laeviceps honey to E. coli was 6 mm, which is smaller than our results. Suntiparapop et al. (2012) investigated the diameter of the inhibition zone of T. laeviceps honey to E. coli and P. aeruginosa found the diameters were 11 and 12 mm, respectively. Their results demonstrate a slightly larger clear zone than our results. The inhibition zone of four stingless bee species honey to gram-positive bacteria were L. flavibasis inhibiting two species of bacteria (B. subtilis and C. botulinum), L. doipaensis honey inhibiting two species of bacteria (C. botulinum and S. aureus). The inhibition zone of T. testaceitarsis honey to C. botulinum, S. aureus and S. epidermis were shown. T. laeviceps species complex honey inhibited the three species of gram-positive bacteria, B. subtilis, C. botulinum and S. epidermis. Nevertheless, no inhibition zone was shown for L. furva honey. A. mellifera honey exhibits a clear zone only for 2 species of gram-negative bacteria: E. aerogenesis and E. coli.

The most significant inhibition to C. botulinum was found with L. doipaensis honey (diameter of 22.5 mm). A. mellifera honey shows inhibition properties to C. botulinum and S. epidermis bacteria with a lower inhibition zone diameter than two samples of L. flavibasis honey, three samples of L. doipaensis honey and a sample of T. testaceitarsis honey. A previous study of antimicrobial activity of stingless bee honey from Thailand was reported by Suntiparapop et al. (2012), with the diameter of clear zone of T. laeviceps honey to S. aureus being less than 10 mm, which in contrast to our analysis, demonstrated no inhibition zone found for T. laeviceps honey. However, this species' honey displayed the greatest clear zone (19.6 mm) to S. epidermis bacteria which consistent to two samples of T. laeviceps honey. Chanchao (2013) and Chanchao (2009) examined the diameter of the inhibition zone of T. laeviceps honey to S. aureus it was found that the clear zone was 14 mm and 38 mm, respectively (the clear zone diameter measurement was done after 72 h following applying honey into culture) which is dissimilar to our results. We found no inhibition zone of T. laeviceps honey to S. aureus.

Stingless bee honey from four species of stingless bee demonstrated antimicrobial activity against the bacteria strains Minimum Inhibitory (MIC) and Minimum Bactericidal Concentration (MBC) as shown in Table 3. The MIC and MBC showed the inhibition was in the range of 3.12 to 50% and 3.12 to 100%, respectively. L. flavibasis honey exhibited activity against two species of gram-positive bacteria (B. subtilis and C. botulinum). L. doipaensis honey showed inhibition to C. botulinum and S. aureus. T. testaceitarsis complex and T. laeviceps honey displayed activity against three species of bacteria except for B. subtilis for T. testaceitarsis complex honey and S. aureus for T. laeviceps complex honey. Our results of four species of stingless bee honey showed the activities against gram-negative bacteria in different concentrations. Suntiparapop et al. (2012) examined the MIC of stingless bee honey from one species of stingless bee (T. laeviceps). The lowest concentration found against bacteria (Micrococcus luteus) was 7.03%. For S. aureus bacteria, their result showed a concentration of 11.76%, which is dissimilar to our result which displays no inhibition of S. aureus from the same species of honey. However, our MIC result of S. epidermis was 3.12% lower than the reported 8.0% of Suntiparapop et al. (2012).

Khongkwanmueang et al., (2020) reveal the antibacterial properties of honey from the T. laeviceps species complex from eastern Thailand (Rayong, Chanthaburi, and Trat provinces), was more susceptible to Gram-negative bacteria than Gram-positive bacteria. The MIC values of honey samples against four bacteria strains were ranged from 10 – 30%, the lowest MIC value was 10% against S. typhimurium, and the lowest MBC value was 25% against B. cereus, S. aureus, and S. typhimurium. In contrast, in our study, the MIC values of honey samples, particularly from T. laeviceps, ranged from 3.12 - 50%, and the lowest MIC values was 25% against S. typhimurium, and the lowest MBC value was 3.12% against B. cereus.

The comparison of antibacterial activity against human pathogenic bacteria between honey samples from different species and geographical origins is shown in Figure 1. Overall, the Gram-negative bacteria were more sensitive to stingless bee honey samples than Gram-negative bacteria as per the average MIC and MBC values. For example, the honey of L. doipaensis effectively inhibited Gram-negative bacteria with the lowest MIC and MBC value (3.12 and 6.25%v/v, respectively) compared with other honey samples. For Gram-positive bacteria, all four honey samples were effective in inhibiting C. botulinum (MBC values ranging from 25 to 3.12%v/v). Several studies report that S. aureus was more susceptible to the bacterial inhibition of stingless bee honey than E. coli (Brown et al., 2020; Omar et al., 2019).

Table 2. Antibacterial activity of stingless bee honey compared to A. mellifera honey (agar disc-diffusion method)

Bee Species	Diameter of Inhibition Zone in mm								
	Gram-positive				Gram-negative				
	B. subtilis	*C. botulinum*	*S. aureus*	*S. epidermis*	*E. aerogenes*	*E. coli*	*P. aeruginosa*	*S. choleraesuis*	*S. typhimurium*
L. flavibasis	6.9± 0.2	17.0[a] ± 2.0	-	-	12.2[a] ± 1.3	14.0[a] ± 1.0	13.7[a] ± 2.4	10.3± 1.5	13.0± 2.1
L. doipaensis	-	22.5[a] ± 0.0	6.9[b] ± 0.2	-	13.2[a] ± 2.8	10.3[b] ± 1.0	12.2[a] ± 1.0	7.8± 0.8	8.2± 0.6
T. testaceitarsis	-	10.5[b] ± 0.5	12.0[a] ± 1.0	13.0[b] ± 1.0	7.8[b] ± 1.1	6.5[b] ± 0.0	7.5[b] ± 0.0	8.0± 0.0	10.7± 0.2
T. laeviceps complex	7.7± 0.3	10.2[b] ± 0.8		18.3[a] ± 3.2	10.7[a] ± 1.6	9.5[b] ± 2.3	8.1[b] ± 0.5	8.5± 1.0	8.7± 1.2
L. furva	-	-	-	-	-	-	-	-	-
A. mellifera	-	12.5[b] ± 0.5	-	9.5[b] ± 0.5	12.5[a] ± 0.5	14.0[a] ± 1.0	-	-	-

However, in our study, E. coli and Gram-negative bacteria were more sensitive to the antibacterial action of stingless bee honey. These results agreed with other studies which have shown honey exhibited a more significant inhibitory effect on E. coli with a larger inhibition zone and lower MIC/MBC than S. aureus (Ng et al., 2020). Honey was found to target the cell wall and lipopolysaccharide outer membrane of E. coli, causing the cell wall destruction and increased permeability of the outer membrane, leading to cell lysis (Brudzynski & Sjaarda, 2014). The antimicrobial activity of honey is associated with various factors, such as moisture, phenolics and flavonoids content, hydrogen peroxide content, and pH. The isolation and characterization of bioactive compounds in stingless bee honey are still required for better understanding of the significant divergence in stingless bee honey composition and geographical origin (Jibril et al., 2019).

The average of inhibition zone excluded the paper disc (mm) by agar disc-diffusion method with SD values (n=3). The small case letters (a,b and c) are indicate significant differences among honey samples. The data were analyzed using One-Way ANOVA followed by post hoc Turkey test (p<0.05).

Table 3. MIC (%v/v) and MBC (% v/v) of stingless bee honey from four species.

Bacteria	Stingless Bee Species							
	L. flavibasis		*L. doipaensis*		*T. testaceitarsis* Complex		*T. laeviceps* Complex	
	MIC	MBC	MIC	MBC	MIC	MBC	MIC	MBC
Gram positive bacteria								
B. subtilis	25	50	-	-	-	-	3.12	3.12
C. botulinum	12.5	25	3.12	3.12	3.12	6.25	12.5	12.5
S. aureus	-	-	25	50	12.5	25	-	-
S. epidermis	-	-	-	-	12.5	25	3.12	6.35
Gram negative bacteria								
E. aerogenes	3.12	6.25	3.12	6.25	3.12	6.25	50	100
E. coli	3.12	6.25	3.12	6.25	12.5	25	25	50
P. aeruginosa	12.5	25	3.12	6.25	6.25	6.25	50	50
S. choleraesuis	6.25	6.25	6.25	6.25	12.5	25	50	50
S. typhimurium	3.12	6.25	6.25	6.25	3.12	6.25	25	50

Figure 1. Comparison of the MIC and MBC value (%v/v) among stingless bee samples from four species against the pathogenic bacteria.

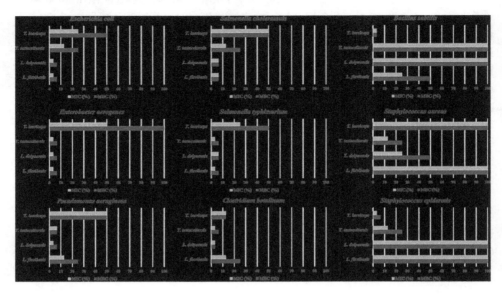

ACKNOWLEDGMENT

Any acknowledgment to fellow researchers or funding grants should be placed within this section.

REFERENCES

Alimentarius, C. (2001). Revised codex standard for honey. *Codex Stan, 12*, 1982.

Alvarez-Suarez, J. M., Tulipani, S., Díaz, D., Estevez, Y., Romandini, S., Giampieri, F., Damiani, E., Astolfi, P., Bompadre, S., & Battino, M. (2010). Antioxidant and antimicrobial capacity of several monofloral Cuban honeys and their correlation with color, polyphenol content and other chemical compounds. *Food and Chemical Toxicology, 48*(8–9), 2490–2499. doi:10.1016/j.fct.2010.06.021 PMID:20558231

Biluca, F. C., Braghini, F., Gonzaga, L. V., Costa, A. C. O., & Fett, R. (2016). Physicochemical profiles, minerals and bioactive compounds of stingless bee honey (Meliponinae). *Journal of Food Composition and Analysis, 50*, 61–69. doi:10.1016/j.jfca.2016.05.007

Bogdanov, S. (1997). Nature and origin of the antibacterial substances in honey. *Lebensmittel-Wissenschaft + Technologie*, *30*(7), 748–753. doi:10.1006/fstl.1997.0259

Bogdanov, S., Jurendic, T., Sieber, R., & Gallmann, P. (2008). Honey for nutrition and health: A review. *Journal of the American College of Nutrition*, *27*(6), 677–689. doi:10.1080/07315724.2008.10719745 PMID:19155427

Brown, E., O'Brien, M., Georges, K., & Suepaul, S. (2020). Physical characteristics and antimicrobial properties of Apis mellifera, Frieseomelitta nigra and Melipona favosa bee honeys from apiaries in Trinidad and Tobago. *BMC Complementary Medicine and Therapies*, *20*(1), 1–9. doi:10.118612906-020-2829-5 PMID:32178659

Brudzynski, K., & Sjaarda, C. (2014). Antibacterial compounds of Canadian honeys target bacterial cell wall inducing phenotype changes, growth inhibition and cell lysis that resemble action of β-lactam antibiotics. *PLoS One*, *9*(9), 1–11. doi:10.1371/journal.pone.0106967 PMID:25191847

Chanchao, C. (2009). Antimicrobial activity by Trigona laeviceps (stingless bee) honey from Thailand. *Pakistan Journal of Medical Sciences*, *25*(3), 364–369.

Chanchao, C. (2013). Bioactivity of Honey and Propolis of Tetragonula laeviceps in Thailand. In *Pot-Honey* (pp. 495–505). Springer. doi:10.1007/978-1-4614-4960-7_36

Cho, K. S., & Tony Ng, H. K. (2021). Tolerance intervals in statistical software and robustness under model misspecification. *Journal of Statistical Distributions and Applications*, *8*(1), 1–49. doi:10.118640488-021-00123-2

Chuttong, B., Chanbang, Y., & Burgett, M. (2014). Meliponiculture: Stingless bee beekeeping in Thailand. *Bee World*, *91*(2), 41–45. doi:10.1080/000577 2X.2014.11417595

Chuttong, B., Chanbang, Y., Sringarm, K., & Burgett, M. (2016). Physicochemical profiles of stingless bee (Apidae: Meliponini) honey from South east Asia (Thailand). *Food Chemistry*, *192*, 149–155. doi:10.1016/j.foodchem.2015.06.089 PMID:26304332

Cortopassi-Laurino, M., Imperatriz-Fonseca, V. L., Roubik, D. W., Dollin, A., Heard, T., Aguilar, I., Venturieri, G. C., Eardley, C., & Nogueira-Neto, P. (2006). Global meliponiculture: Challenges and opportunities. *Apidologie*, *37*(2), 275–292. doi:10.1051/apido:2006027

Dajanta, K., Chukeatirote, E., & Apichartsrangkoon, A. (2012). Nutritional and physicochemical qualities of Thua Nao (Thai traditional fermented soybean). *Warasan Khana Witthayasat Maha Witthayalai Chiang Mai, 39*(4), 562–574.

Escuredo, O., Silva, L. R., Valentão, P., Seijo, M. C., & Andrade, P. B. (2012). Assessing Rubus honey value: Pollen and phenolic compounds content and antibacterial capacity. *Food Chemistry, 130*(3), 671–678. doi:10.1016/j.foodchem.2011.07.107

Guerrini, A., Bruni, R., Maietti, S., Poli, F., Rossi, D., Paganetto, G., Muzzoli, M., Scalvenzi, L., & Sacchetti, G. (2009). Ecuadorian stingless bee (Meliponinae) honey: A chemical and functional profile of an ancient health product. *Food Chemistry, 114*(4), 1413–1420. doi:10.1016/j.foodchem.2008.11.023

Jibril, F. I., Hilmi, A. B. M., & Manivannan, L. (2019). Isolation and characterization of polyphenols in natural honey for the treatment of human diseases. *Bulletin of the National Research Center, 43*(1), 1–9. doi:10.118642269-019-0044-7

Khongkwanmueang, A., Nuyu, A., Straub, L., & Maitip, J. (2020). Physicochemical profiles, antioxidant and antibacterial capacity of honey from stingless bee tetragonula laeviceps species complex. *E3S Web of Conferences, 141*, 3007.

Kwakman, P. H. S., te Velde, A. A., de Boer, L., Speijer, D., Christina Vandenbroucke-Grauls, M. J., & Zaat, S. A. J. (2010). How honey kills bacteria. *The FASEB Journal, 24*(7), 2576–2582. doi:10.1096/fj.09-150789 PMID:20228250

Lusby, P. E., Coombes, A. L., & Wilkinson, J. M. (2005). Bactericidal activity of different honeys against pathogenic bacteria. *Archives of Medical Research, 36*(5), 464–467. doi:10.1016/j.arcmed.2005.03.038 PMID:16099322

Maitip, J., Mookhploy, W., Khorndork, S., & Chantawannakul, P. (2021). Comparative Study of Antimicrobial Properties of Bee Venom Extracts and Melittins of Honey Bees. *Antibiotics (Basel, Switzerland), 10*(12), 1–14. doi:10.3390/antibiotics10121503 PMID:34943715

McLoone, P., Warnock, M., & Fyfe, L. (2016). Honey: A realistic antimicrobial for disorders of the skin. *Journal of Microbiology, Immunology, and Infection, 49*(2), 161–167. doi:10.1016/j.jmii.2015.01.009 PMID:25732699

Michener, C. D. (2013). The meliponini. In *Pot-honey* (pp. 3–17). Springer. doi:10.1007/978-1-4614-4960-7_1

Molan, P. C. (1992). The antibacterial activity of honey: 1. The nature of the antibacterial activity. *Bee World, 73*(1), 5–28. doi:10.1080/0005772X.1992.11099109

Ng, W.-J., Sit, N.-W., Ooi, P. A.-C., Ee, K.-Y., & Lim, T.-M. (2020). The antibacterial potential of honeydew honey produced by stingless bee (Heterotrigona itama) against antibiotic resistant bacteria. *Antibiotics (Basel, Switzerland)*, *9*(12), 1–16. doi:10.3390/antibiotics9120871 PMID:33291356

Nombré, I., Schweitzer, P., Boussim, J. I., & Rasolodimby, J. M. (2010). Impacts of storage conditions on physicochemical characteristics of honey samples from Burkina Faso. *African Journal of Food Science*, *4*(7), 458–463.

Omar, S., Mat-Kamir, N. F., & Sanny, M. (2019). Antibacterial activity of Malaysian produced stingless-bee honey on wound pathogens. *Journal of Sustainability Science and Management*, *14*, 67–79.

Pauly, A., Pedro, S. R. M., Rasmussen, C., & Roubik, D. W. (2013). Stingless bees (Hymenoptera: Apoidea: Meliponini) of French Guiana. In Pot-Honey (pp. 87–97). Springer.

Rasmussen, C. (2008). Catalog of the Indo-Malayan/Australasian stingless bees (Hymenoptera: Apidae: Meliponini). *Zootaxa*, *1935*(1), 1–80. doi:10.11646/zootaxa.1935.1.1

Souza, B., Roubik, D., Barth, O., Heard, T., Enríquez, E., Carvalho, C., Villas-Bôas, J., Marchini, L., Locatelli, J., & Persano-Oddo, L. (2006). Composition of stingless bee honey: Setting quality standards. *Interciencia*, *31*(12), 867–875.

Suntiparapop, K., Prapaipong, P., & Chantawannakul, P. (2012). Chemical and biological properties of honey from Thai stingless bee (Tetragonula leaviceps). *Journal of Apicultural Research*, *51*(1), 45–52. doi:10.3896/IBRA.1.51.1.06

Vallianou, N. G., Gounari, P., Skourtis, A., Panagos, J., & Kazazis, C. (2014). Honey and its anti-inflammatory, anti-bacterial and anti-oxidant properties. *General Medicine (Los Angeles, Calif.)*, *2*(132), 1–5. doi:10.4172/2327-5146.1000132

Vit, P., Pedro, S. R. M., & Roubik, D. (2013). Pot-honey: A legacy of stingless bees. Springer Science & Business Media.

Vit, P., Bogdanov, S., & Kilchenmann, V. (1994). Composition of Venezuelan honeys from stingless bees (Apidae: Meliponinae) and Apis mellifera L. *Apidologie*, *25*(3), 278–288. doi:10.1051/apido:19940302

White, J. W., & Doner, L. W. (1980). Honey composition and properties. *Beekeeping in the United States Agriculture Handbook*, *335*, 82–91.

Wikler, M. A. (2006). Methods for dilution antimicrobial susceptibility tests for bacteria that grow aerobically: Approved standard. *Clsi (Nccls)*, *26*, M7–A7.

Zamora, G., Arias, M. L., Aguilar, I., & Umaña, E. (2013). Costa Rican pot-honey: its medicinal use and antibacterial effect. In *Pot-Honey* (pp. 507–512). Springer. doi:10.1007/978-1-4614-4960-7_37

Chapter 4
Approach for the Domestication and Propagation of Stingless Bees

Ali Agus
Universitas Gadjah Mada, Indonesia

Agussalim Agussalim
Faculty of Animal Science, Universitas Gadjah Mada, Indonesia

ABSTRACT

Domestication and propagation in stingless bees is called meliponiculture. The aims of meliponiculture are to make it easier to control the colonies health and development and to make it easy when harvesting stingless bee products (honey, bee bread, and propolis), furthermore, for advanced study and development like multiple colonies, to produce honey, bee bread, and propolis. Therefore, this paper focuswa on stingless bees Tetragonula laeviceps: the domestication and propagation technique, production of stingless bee products (honey, bee bread, and propolis), the daily activity of workers (foragers), the chemical composition (glucose, fructose, sucrose, reducing sugar, moisture, protein, ash, phenolic, flavonoid, vitamin C, antioxidant activity, minerals content, and amino acids) of honey from T. laeviceps, the pests and the challenges in meliponiculture of stingless bees.

INTRODUCTION

The stingless bees are eusocial bees and include the tropical group, whereas the number of species have been identified are 500 species and possibly more than 100 species yet undescribed. The stingless bees include family Apidae, sub-family Apinae, and

DOI: 10.4018/978-1-6684-6265-2.ch004

tribe Meliponini. In tropical regions of the world, they are divided into three zones are the American tropics (Neotropics), sub-Saharan African (Afrotropical region), and the Indoaustralian (Austroasian) region (Michener, 2007; 2013). In Indonesia, the stingless bees species at least 46 species from genus *Austroplebeia Moure, Geniotrigona Moure, Heterotrigona Schwarz, Homotrigona Moure, Lepidotrigona Schwarz, Lisotrigona Moure, Papuatrigona Michener* and *Sakagami, Pariotrigona Moure, Tetragonula Moure, and Wallacetrigona Engel* and *Rasmussen* which are spread in several islands like Sumatera, Java, Timor, Borneo, Sulawesi, Ambon, Maluku, and Irian Jaya (Papua) (Kahono et al., 2018).

Stingless bees species in Indonesia are *Austroplebeia cincta, Geniotrigona lacteifasciata, G. thoracica, Heterotrigona erythrogastra, H. itama, Platytrigona flaviventris, P. hobbyi, P. keyensis, P. lamingtonia, P. planifrons, Sahulotrigona atricornis, Sundatrigona lieftincki, S. moorei, Homotrigona aliceae, H. anamitica, H. fimbriata, Lophotrigona canifrons, Odontotrigona haematoptera, Tetrigona apicalis, T. binghami, T. vidua, Lepidotrigona javanica, L. latebalteata, L. nitidiventris, L. terminata, L. trochanterica, L. ventralis, Lisotrigona cacciae, Papuatrigona genalis, Pariotrigona pendleburyi, Tetragonila atripes, T. collina, T. fuscibasis, Tetragonula biroi, T. clypearis, T. drescheri, T. fuscobalteata, T. geissleri, T. laeviceps, T. melanocephala, T. melina, T. minangkabau, T. reepeni, T. sapiens, T. sarawakensis,* and *Wallacetrigona incisa* (Kahono et al., 2018). In addition, the recently reported that the seven species of stingless bees found in the Special Region of Yogyakarta are *T. laeviceps, T. biroi, T. sapiens, T. iridipennis, T. sarawakensis, Lepidotrigona terminata,* and *Heterotrigona itama* (Trianto & Purwanto, 2020).

Furthermore, the *T. laeviceps* is a species whose distribution in Indonesia is quite large, whereas all the islands are covered (Kahono et al., 2018). In Indonesia, stingless bees can be found in the forest, plantations, and they create the nest in wood or tree trunk, bamboo, sugar palm stalks, and in the ground. Especially stingless bee *Tetragonula laeviceps* can be found nesting in bamboo, wood or tree trunk, and in the ground (Agus et al., 2019a,b; Agusalim et al., 2019a; 2015). Therefore, this paper is focused on stingless bee *T. laeviceps*: the domestication and propagation technique, production of stingless bee products, the daily activity of workers (foragers), the chemical composition of honey, the pests, and the challenges in meliponiculture.

DOMESTICATION AND PROPAGATION TECHNIQUE

Domestication in our research focused on the stingless bee *T. laeviceps* and the natural hive is obtained from the bamboo hives from the plantation in Yogyakarta, Indonesia and the domestication process can be shown in Figure 1.

Figure 1. The process of domestication of T. laeviceps
Note: a. Tetragonula laeviceps in the bamboo hive; b. Description of the colony in the bamboo hive; c. The brood cells, queen, workers, and drones have been moved from bamboo hives to box hives; d. The house of bee to placed box hives for meliponiculture.

The domestication process consists of the colonies from bamboo being split and then the brood cells, queen bee, drones, and workers are moved from bamboo to box hives. Afterward, the colonies are placed in bee house (Figure 2) for meliponiculture. The box hives used in our study have a size 35 x 15 x 12 cm for length, width, and height, respectively and are made from the dry board with a thickness of 2 cm. Several factors that are very important after domestication are to avoid the colonies from the ants, lizards, and spiders. When the colonies are domesticated, must be performed carefully to avoid dripping honey in the hives because it can be inviting the ants. Consequently, the colonies in the hives can be by ants if they can be an entrance to hives, especially the new colony after domestication from bamboo hive to box hive. In nature, the domestication in stingless bees can be performed with three methods are the tree cutting, grafting, and trap system (Buchori et al., 2022). The colonies from stingless bee *T. laeviceps* in Yogyakarta are found mostly nesting in bamboo hives. The colony condition of stingless bee of *T. laeviceps* after four months of meliponiculture in Faculty of Animal Science, Universitas Gadjah Mada (UGM), Indonesia is shown in Figure 2.

Figure 2. The colony condition of stingless bee T. laeviceps
Note: The yellow circle is pots honey and the red circle is brood cells

Figure 2 includes a good colony of stingless bee *T. laeviceps* after meliponiculture for four months is characterized by brood cells and honey pots which are much. In addition, the forages as the source of nectar to produce honey in the Faculty of Animal Science UGM consist of calliandra, bananas, mangoes, chicory, sunflowers, tamarind, catappa, indigofera, kapok, syzygium, alfalfa, kepel, star fruit, bilimbi, matoa, water apple, guava, lemon, gliricidia, chili, caimito, rambutan, and canarium (Agussalim et al., 2020).

PRODUCTION OF STINGLESS BEE PRODUCTS

In general, stingless bees can be producing honey, bee bread, and propolis. Production of honey from stingless bees is lower compared with honey production of honeybees from genus Apis like *Apis mellifera, A. cerana, A. dorsata*, etc. In addition, propolis production in stingless bees is higher than in honeybees. *Tetragonula laeviceps* is one of the species of stingless bee and more meliponiculture by beekeepers in Indonesia. Production of honey, bee bread, and propolis from stingless bee *T. laeviceps* in each region in Indonesia for meliponiculture is varied. Honey from stingless bee *T. laeviceps* is made by workers in the hive required 3 to 4 days after domestication (from bamboo hive to box hive) to produce 2 to 3 pots that are filled with honey. Production of honey *T. laeviceps* meliponiculture in Faculty of Animal Science UGM (Yogyakarta, Indonesia) is ranged from 60 to 263 ml/colony (79.2 to 328 g) with the mean 176 ml (223 g) after four months meliponiculture (Agussalim et al., 2020), 49.20 to 66.60 ml/colony after two months meliponiculture for *Tetragonula* sp. in Gunungkidul (Yogyakarta) (Agussalim et al., 2017), 1.44 g/colony/month for Cibodas, 0.93 g/colony/month for Cileunyi Wetan (Bandung) from *T. laeviceps* (Abduh et al., 2020).

The different production of honey from each region for meliponiculture is caused by the different regions as the location to meliponiculture, different plant types as the nectar source to produce honey, the different environment conditions (temperature, humidity, and altitude). In addition, the foragers activity especially when collecting nectar from plant flowers (floral nectar) and extrafloral nectar, the distance of meliponiculture site to plant types that impact the number of nectars can be collected by foragers of stingless bees. Production of propolis from stingless bee *T. laeviceps* is varied from each region for meliponiculture depending on the availability of plants as the resin sources and the active level of foragers to collect resin much more. Propolis production from stingless bee *T. laeviceps* that meliponiculture in Faculty of Animal Science UGM (Yogyakarta, Indonesia) is ranged from 15.4 to 77.2 g/colony obtained from honey pots after meliponiculture for four months (Agussalim et al., 2020), 18.20 to 30.80 g/colony after two months of meliponiculture for propolis from honey pots for *Tetragonula* sp. (Agussalim et al., 2015), total production of propolis is 4.26 g/colony/month (propolis from the frame is 3.80 g/colony/month and 0.46 g/colony/month from honey pots) in Cileunyi Wetan region for meliponiculture, and 4.54 g/colony/month (propolis from the frame is 4.24 g/colony/month, and 0.30 g/colony/month from honey pots) in Cibodas (Bandung) (Abduh et al., 2020). The different production of propolis in each region in Indonesia is related to the availability of resin from living plants, the number of workers especially foragers that collect resin, the productivity of queen bee to produce eggs as the workers.

Production of bee bread from stingless bee *Tetragonula* sp. is ranged from 1.02 to 4.56 g/colony after two months of meliponiculture using various box hives in Gunungkidul (Yogyakarta, Indonesia) (Agus et al., 2019a). Production of bee bread is dependent on the availability of pollen from plant flowers, the exit activity of foragers to collect pollen, the population of workers especially foragers, and region condition (temperature and humidity). Bee bread is the main source of protein in the hive which is required by workers bee to produce royal jelly as the feed of queen bee, impact the increase of productivity of queen bee to produce eggs as the workers.

DAILY ACTIVITIES OF WORKERS (FORAGERS)

The activities of workers in the hives are to produce stingless bee products (honey, bee bread, propolis, royal jelly), to carrying and feeding (eggs, larvae, and queen), while out of the hives to collect nectar, pollen, resin, water, and some materials needed to construct the nest and hives. This paper focuses on the entrance and exit activities of foragers of stingless bee *T. laeviceps*. The daily activity of stingless bee *T. laeviceps* in the morning (07.00 to 11.00 am) is higher than activity in the afternoon (2.00 to 5.00 pm). The number of foragers who bring in pollen to the hive in the morning (07.00 to 11.00 am) is ranged from 20.15 to 25.36 times per 5 minutes per colony, while in the afternoon is ranged from 12.19 to 12.85 times per 5 minutes per colony. Furthermore, the foragers are exit from the hive to collect nectar, pollen, and resin in the morning started is ranged from 05.25 to 05.30 am and the first time bring in pollen to the hive is ranged from 05.40 to 05.45 am (Agus et al., 2019a).

In addition, the exit and entrance of the hive of stingless bee *Tetragonula* sp. from the hive is ranged from 37 to 43 times per 5 minutes per colony and ranged from 38 to 46 times per 5 minutes per colony, respectively (Agussalim et al., 2017). The exit activity of foragers *Tetragonula* sp. in the bamboo hive in the morning (08.00 am) is ranged from 34.7 to 37.5 times and higher than in the afternoon (04.00 pm) is ranged from 24.9 to 25.5 times per 5 minutes per colony. Furthermore, the exit activity of foragers in the box hive is ranged from 49.2 to 51.3 times in the morning is higher than in the afternoon is ranged from 29.0 to 29.6 times per 5 minutes per colony (Erwan et al., 2021). Furthermore, reported that the exit activity of foragers *Tetragonula* sp. in the morning (08.00 am) from the bamboo hive is 36.6 times and in the afternoon is 25.3 times, while in the box hive is 50.1 times in the morning (08.00 am) and 29.3 times in the afternoon. The daily activity of stingless bee *T. laeviceps* in each colony is differed depending on the population of bee in the colony, the environment condition (temperature, humidity, air velocity, and light intensity), and the availability of food (nectar, honeydew, and pollen) (Erwan et al., 2020).

CHEMICAL COMPOSITION OF HONEY

The moisture content is one of the very important parameters which influence the physical properties of honey such as crystallization process, colour, taste, flavour, and solubility (da Silva et al., 2016; Escuredo et al., 2013). The moisture content of honey from stingless bee *T. laeviceps* is ranged from 21.21 to 26.81% (Agussalim et al, 2019) and the moisture content is lower to those previously studied (Biluca et al., 2016; Guerrini et al., 2009; Ranneh et al., 2018; Souza et al., 2006; Suntiparapop et al., 2012). The moisture content of stingless bee honey is high because they also collect some material from ripe fruit which is higher in moisture content. In addition, the stingless bees species have not developed a mechanism of behaviour to evaporate water compared with honeybees from the genus Apis (Suntiparapop et al., 2012). The moisture is acceptable by Indonesian standard for stingless bee honey (SNI, 2018), but the international standard by Codex Alimentarius has not been set for honey from stingless bees.

The ash content of honey from stingless bee *T. laeviceps* is ranged from 0.07 to 0.49 g/100 g of honey (Agussalim et al., 2019) and contains the minerals consist of

Ca is ranged from 2,964.86 to 3,256.83 ppm, Cu 3.9 to 12.2 ppm, Fe 9.75 to 45.4 ppm, Mg 289.48 to 2,389.65 ppm, Mn 12.5 to 33.9 ppm, Na 234.90 to 1,142.87 ppm, K 2,498.65 to 20,110.80 ppm, Zn 12.0 to 23.5 ppm, and Al 858.89 to 1,120.71 ppm. The abundance of minerals of honey from *T. laeviceps* is potassium followed by calcium (Sabir et al., 2021). The ash content of honey is related to mineral content present in the honey and indicates the geographical origin marker (da Silva et al., 2016; Suntiparapop et al., 2012). The ash and mineral content of honey is affected by the soil nutrient and the nectar source from each plant flower (da Silva et al., 2016; Karabagias et al., 2014). In addition, the ash and mineral content of honey related to colour and flavour, where the higher mineral content characterized by honey is darker and stronger flavour if compared with honey bright colour with the low mineral content (da Silva et al., 2016; Escuredo et al., 2013; Karabagias et al., 2014).

The protein content of honey from stingless bee *T. laeviceps* is ranged from 0.18 to 0.72 g/100 g of honey (Agussalim et al., 2019), while the amino acids present in honey from *T. laeviceps* consist of arginine is 591.83 mg/kg, histidine 561.93 mg/kg, lysine 882.03 mg/kg, phenylalanine 232.74 mg/kg, isoleucine 12.34 mg/kg, leucine 73.55 mg/kg, methionine 0.29 mg/kg, valine 20.39 mg/kg, threonine 45.72 mg/kg, tyrosine 9.24 mg/kg, proline 60.56 mg/kg, glutamic acid 119.82 mg/kg, aspartic acid 77.31 mg/kg, serine 168.65 mg/kg, alanine 62.46 mg/kg, and glycine 60.26 mg/kg of honey (Agussalim et al., 2019b). The protein and amino content in honey depend on the nectar source from plant flowers and the pollen source from plant flowers. The main source of protein for honeybees or stingless bees is pollen (da Silva et al., 2016; Karabagias et al., 2014). In honey, the amino acid proline can be used to detect the adulteration of honey which is made from sugar like cane, coconut, and palm sugars. The minimum content of proline is 180 mg/kg is acceptable for the minimum proline content for pure honey especially in honey from Apis mellifera (da Silva et al., 2016), but in stingless bees honey has not been confirmed as one of the parameters to detect the adulteration of honey. The lower content of proline amino acid is related to not complete maturity of honey and might be an adulteration of honey, but it needed the advanced study to verify this case.

The sugar content of honey from stingless bee *T. laeviceps* is lower than honey from Apis genus like *Apis mellifera, Apis cerana, Apis dorsata*, etc. The different geographical origins for meliponiculture impacted the sugars content of honey. The sugar content of honey from stingless bee *T. laeviceps* consist of glucose is ranged from 11.49 to 22.78% (w/w), fructose 7.79 to 22.92% (w/w), sucrose 2.56 to 4.49% (w/w), and reducing sugar 44.07 to 60.14 g/100 g of honey, the total of glucose and fructose 30.57 to 43.16% (w/w), and the ratio of fructose/glucose 0.34 to 1.99 (Agussalim et al., 2019a). The sugars content of honey is affected by the different nectar source (plant types) to produce honey, geographical origin for meliponiculture, climate (temperature and humidity), processing (heated, manipulation, and packaging), and storage time (da Silva et al., 2016; Escuredo et al., 2014; Tornuk et al., 2013). The sugars content in our study is differs from those previously studied (Biluca et al., 2016; Chuttong et al., 2016; Guerrini et al., 2009; Oddo et al., 2008; Souza et al., 2006; Suntiparapop et al., 2012). The sucrose content of honey is one of the very important parameters to evaluate the adulteration or manipulation of honey and the maturity of honey. The high sucrose content from honey may indicate manipulation and adulteration of honey using sugar (sugar cane, refined beet sugar, sucrose syrups). In addition, it also indicates the early harvest of honey so the sucrose is not completely transformed into glucose and fructose (da Silva et al., 2016; Escuredo et al., 2013; Puscas et al., 2013; Tornuk et al., 2013).

The bioactive compounds of honey from stingless bee *T. laeviceps* have been studied to consist of total phenolic and total flavonoid. In addition, vitamin C and DPPH antioxidant activity were also studied. The total phenolic content of honey from stingless bee *T. laeviceps* is ranged from 0.54 to 1.69% GAE (w/w), total flavonoid content 0.21 to 0.90 mg QE/g, vitamin C content 5.67 to 7.88 mg/100 g, and the DPPH antioxidant activity 47.3 to 91.2% at concentration 0.1 mM (Agus et al., 2019b). The total phenolic content is differs from those previously studied (Biluca et al., 2016; Guerrini et al., 2009; Oddo et al., 2008; Ranneh et al., 2018), total flavonoid content (Oddo et al., 2008; Ranneh et al., 2018), vitamin C (Ranneh et al 2018), and DPPH antioxidant activity (Guerrini et al., 2009; Oddo et al., 2008; Ranneh et al., 2018). The phenolic, flavonoid, and vitamin C is affected by the nectar source (plant types), while antioxidant activity is affected by the metabolites secondary content such as phenolic and flavonoid. Based on the Pearson correlation coefficients in our study that the DPPH antioxidant activity is affected by the total phenolic and total flavonoid contents (Agus et al., 2019b).

PESTS AND CHALLENGES IN MELIPONICULTURE

The pest in meliponiculture of stingless bee *T. laeviceps* in Yogyakarta consists of ants, spiders, beetles, lizards, and black soldier fly. Furthermore, reported that the pest in meliponiculture of stingless bees consists of Drosophila sp., beetles from family Nitidulidae, ants, honey bear, spiders (*Argiope versicolor*), lizards (*Hemidactylus frenatus* and *H. garnotii*), termite (*Nasutitermes javanicus*), and wasp (*Rhynchium haemorrhoidale*) (Buchori et al., 2022). All these pests can damage the colonies and decrease the production of honey and bee bread. The challenges in meliponiculture of stingless bees consist of the illegal logging of the forest to give the colonies, the high use of pesticide in agricultural and plantation sector that affects to less of bees age or residual in stingless bee products, the limited of sustainable plants as the forages (source of nectar and pollen).

CONCLUSION

Currently, the study about the domestication of stingless bees in Indonesia has been studied by the scientist that collaborates with beekeepers about the production of stingless bee products but needed advanced study like multiple colonies especially queen bee to avoid the illegal logging of the forest. In addition, further is needed collaboration between scientists to study the chemical composition of stingless bees' products from various species and locations to evaluate the quality and compared with the quality of honeybee's products. The challenges of meliponiculture consist of illegal logging of the forest to give the colonies, the high use of pesticide in the agricultural and plantation sector, the limited sustainable plants as the forages (source of nectar and pollen).

ACKNOWLEDGMENT

The authors would like to thank the Directorate of Research and Community Service, Ministry of Education, Culture, Research, and Technology of the Republic of Indonesia for financial support of the research through *Penelitian Terapan Unggulan Perguruan Tinggi* (PTUPT, 2018 – 2020) and *Penelitian Disertasi Doktor* (PDD, 2019 – 2020). Also to the Directorate of Research, Reputation Improvement Team World Class University-Quality Assurance Office of Universitas Gadjah Mada for funding sponsor through Post-Doctoral Program 2021.

REFERENCES

Abduh, M. Y., Adam, A., Fadhlullah, M., Putra, R. E., & Manurung, R. (2020). Production of propolis and honey from *Tetragonula laeviceps* cultivated in Modular Tetragonula Hives. *Heliyon*, *6*(11), 1–8. doi:10.1016/j.heliyon.2020.e05405 PMID:33204881

Agus, A., Agussalim, N., Umami, N., & Budisatria, I. G. S. (2019b). Evaluation of antioxidant activity, phenolic, flavonoid and Vitamin C content of several honeys produced by the Indonesian stingless bee: *Tetragonula laeviceps*. *Livestock Research for Rural Development, 31*(10).

Agus, A., Agussalim, A., Umami, N., & Budisatria, I. G. S. (2019a). Effect of different beehives size and daily activity of stingless bee *Tetragonula laeviceps* on bee-pollen production. *Buletin Peternakan*, *43*(4), 242–246. doi:10.21059/buletinpeternak.v43i4.47865

Agussalim, A. A., Umami, N., & Budisatria, I. G. S. (2017). The Effect of Daily Activities Stingless Bees of Trigona sp. on Honey Production. *The 7th International Seminar on Tropical Animal Production*, 223–227.

Agussalim, A. A., Nurliyani, & Umami, N. (2019a). The sugar content profile of honey produced by the Indonesian stingless bee, *Tetragonula laeviceps*, from different regions. *Livestock Research for Rural Development, 31*(6). http://www.lrrd.org/lrrd31/6/aguss31091.html

Agussalim, A. A., Nurliyani, & Umami, N. (2019b). Free amino acids profile of honey produced by the Indonesian stingless bee: *Tetragonula laeviceps. The 8th International Seminar on Tropical Animal Production*, 149–152.

Agussalim, A., Agus, A., Nurliyani, Umami, N., & Budisatria, I. G. S. (2019). Physicochemical properties of honey produced by the Indonesian stingless bee: *Tetragonula laeviceps. IOP Conference Series. Earth and Environmental Science, 387*(1), 012084. Advance online publication. doi:10.1088/1755-1315/387/1/012084

Agussalim, N., Umami, N., & Agus, A. (2020). The honey and propolis production from Indonesian stingless bee: *Tetragonula laeviceps. Livestock Research for Rural Development, 32*(8).

Agussalim, U. N., & Erwan. (2015). Production of Stingless Bees (Trigona sp.) Propolis in Various Bee Hives Design. *The 6th International Seminar on Tropical Animal Production*, 335–338.

Biluca, F. C., Braghini, F., Gonzaga, L. V., Costa, A. C. O., & Fett, R. (2016). Physicochemical profiles, minerals and bioactive compounds of stingless bee honey (Meliponinae). *Journal of Food Composition and Analysis, 50*, 61–69. doi:10.1016/j.jfca.2016.05.007

Buchori, D., Rizali, A., Priawandiputra, W., Raffiudin, R., Sartiami, D., Pujiastuti, Y., Pradana, M. G., Meilin, A., Leatemia, J. A., & Sudiarta, I. P. (2022). Beekeeping and managed bee diversity in Indonesia: Perspective and preference of beekeepers. *Diversity (Basel), 14*(1), 1–14. doi:10.3390/d14010052

Chuttong, B., Chanbang, Y., Sringarm, K., & Burgett, M. (2016). Physicochemical profiles of stingless bee (Apidae: Meliponini) honey from South east Asia (Thailand). *Food Chemistry, 192*, 149–155. doi:10.1016/j.foodchem.2015.06.089 PMID:26304332

da Silva, P. M., Gauche, C., Gonzaga, L. V., Costa, A. C. O., & Fett, R. (2016). Honey: Chemical composition, stability and authenticity. *Food Chemistry, 196*, 309–323. doi:10.1016/j.foodchem.2015.09.051 PMID:26593496

Erwan, A., M., S., Muhsinin, M., & Agussalim. (2020). The effect of different beehives on the activity of foragers, honey potsnumber and honey production from stingless bee Tetragonula sp. *Livestock Research for Rural Development, 32*(10).

Erwan, S., Syamsuhaidi, P., D. K., Muhsinin, M., & Agussalim. (2021). *Propolis mixture production and foragers daily activity of stingless bee Tetragonula sp . in bamboo and box hives.* Academic Press.

Escuredo, O., Dobre, I., Fernández-González, M., & Seijo, M. C. (2014). Contribution of botanical origin and sugar composition of honeys on the crystallization phenomenon. *Food Chemistry, 149*, 84–90. doi:10.1016/j.foodchem.2013.10.097 PMID:24295680

Escuredo, O., Míguez, M., Fernández-González, M., & Carmen Seijo, M. (2013). Nutritional value and antioxidant activity of honeys produced in a European Atlantic area. *Food Chemistry, 138*(2-3), 851–856. doi:10.1016/j.foodchem.2012.11.015 PMID:23411187

Guerrini, A., Bruni, R., Maietti, S., Poli, F., Rossi, D., Paganetto, G., Muzzoli, M., Scalvenzi, L., & Sacchetti, G. (2009). Ecuadorian stingless bee (Meliponinae) honey: A chemical and functional profile of an ancient health product. *Food Chemistry, 114*(4), 1413–1420. doi:10.1016/j.foodchem.2008.11.023

Kahono, S., Chantawannakul, P., & Engel, M. S. (2018). Social bees and the current status of beekeeping in Indonesia. In *Asian beekeeping in the 21st century* (pp. 287–306). Springer. doi:10.1007/978-981-10-8222-1_13

Karabagias, I. K., Badeka, A., Kontakos, S., Karabournioti, S., & Kontominas, M. G. (2014). Characterisation and classification of Greek pine honeys according to their geographical origin based on volatiles, physicochemical parameters and chemometrics. *Food Chemistry, 146*, 548–557. doi:10.1016/j.foodchem.2013.09.105 PMID:24176380

Michener, C. D. (2007). *The bees of the world* (2nd ed.). The Johns Hopkins University Press.

Michener, C. D. (2013). The Meliponini. In Pot - Honey: a Legacy of Stingless Bees (pp. 3–17). Springer. doi:10.1007/978-1-4614-4960-7_1

Oddo, L. P., Heard, T. A., Rodríguez-Malaver, A., Pérez, R. A., Fernández-Muiño, M., Sancho, M. T., Sesta, G., Lusco, L., & Vit, P. (2008). Composition and antioxidant activity of Trigona carbonaria honey from Australia. *Journal of Medicinal Food, 11*(4), 789–794. doi:10.1089/jmf.2007.0724 PMID:19012514

Puscas, A., Hosu, A., & Cimpoiu, C. (2013). Application of a newly developed and validated high-performance thin-layer chromatographic method to control honey adulteration. *Journal of Chromatography. A*, *1272*, 132–135. doi:10.1016/j. chroma.2012.11.064 PMID:23245847

Ranneh, Y., Ali, F., Zarei, M., Akim, A. M., Abd Hamid, H., & Khazaai, H. (2018). Malaysian stingless bee and Tualang honeys: A comparative characterization of total antioxidant capacity and phenolic profile using liquid chromatography-mass spectrometry. *Lebensmittel-Wissenschaft + Technologie*, *89*, 1–9. doi:10.1016/j. lwt.2017.10.020

Sabir, A., Agus, A., Sahlan, M., & Agussalim. (2021). The minerals content of honey from stingless bee *Tetragonula laeviceps* from different regions in Indonesia. *Livestock Research for Rural Development, 33*(2).

SNI. (2018). *Standar nasional Indonesia madu*. Badan Standarisasi Nasional.

Souza, B., Roubik, D., Barth, O., Heard, T., Enríquez, E., Carvalho, C., Villas-Bôas, J., Marchini, L., Locatelli, J., & Persano-Oddo, L. (2006). Composition of stingless bee honey: Setting quality standards. *Interciencia*, *31*(12), 867–875.

Suntiparapop, K., Prapaipong, P., & Chantawannakul, P. (2012). Chemical and biological properties of honey from Thai stingless bee (Tetragonula leaviceps). *Journal of Apicultural Research*, *51*(1), 45–52. doi:10.3896/IBRA.1.51.1.06

Tornuk, F., Karaman, S., Ozturk, I., Toker, O. S., Tastemur, B., Sagdic, O., Dogan, M., & Kayacier, A. (2013). Quality characterization of artisanal and retail Turkish blossom honeys: Determination of physicochemical, microbiological, bioactive properties and aroma profile. *Industrial Crops and Products*, *46*, 124–131. doi:10.1016/j.indcrop.2012.12.042

Trianto, M., & Purwanto, H. (2020). Morphological characteristics and morphometrics of stingless bees (Hymenoptera: Meliponini) in Yogyakarta, Indonesia. *Biodiversitas Journal of Biological Diversity*, *21*(6), 2619–2628. doi:10.13057/biodiv/d210633

Chapter 5
Brazil–Inspired Vertical Hive Technology for the Philippine Version

Leo Grajo
Grajo's Farm, Philippines

ABSTRACT

The Bicol Region is the birthplace of meliponiculture in the Philippines using the native stingless bee species, Tetragonula biroi Friese. Mr. Rodolfo Palconitin of Guinobatan, Albay, started the traditional method of stingless beekeeping using indigenous material, the coconut shell, which he called bao tech or coconut shell technology. It is a form of natural hive duplication wherein coconut shell halves are gradually mounted on top of each other as the colony grows. In this technology, hive product harvesting and colony splitting are done when the stingless bees have filled up the coconut shell halves. Inspired by the visit to the University of Los Baños (UPLB) Bee Program in 2010, the Grajo's Farm started using bao technology with several experimental hives upon return to home.

INTRODUCTION

The Bicol Region is the birthplace of meliponiculture in the Philippines using the native stingless bee species, Tetragonula biroi Friese Belina-Aldemita et al. (2019) of Abante (2020), started the traditional method of stingless beekeeping using indigenous material – the coconut shell, which he called Bao Tech or Coconut shell technology (Fig. 1). It is a form of natural hive duplication wherein coconut shell halves are gradually mounted on top of each other as the colony grows. In this

DOI: 10.4018/978-1-6684-6265-2.ch005

technology, hive product harvesting and colony splitting are done when the stingless bees have filled up the coconut shell halves. Inspired by the visit to the University of Los Baños (Baroga-Barbecho & Cervancia, 2019), Grajo's Farm started using Bao Technology with several experimental hives upon return home. Cervancia (2018), former Director of the UPLB Bee.

The program encouraged them to raise stingless bees instead of Apis mellifera. At the beginning of their stingless beekeeping journey, it was very challenging due to a lack of technical knowledge, mentor, and dependence on bee hunters for feral colony supply. Some of the feral colonies bought were damaged, while others were drowned in honey, resulting in robbing and infestation. This incident has led to the colony collapse of their stingless bees. But these trials did not discourage them from continuing but motivated them more instead.

Three years later, they came across the name of Dr. Giorgio Venturieri while doing online research on meliponiculture, specifically hive design. Dr. Venturieri was a senior research scientist of Empresa Brasileira de Pesquisa Agropecuria (EMBRAPA) or the Brazilian Agricultural Research Corporation who spent his sabbatical years in Australia (Fig. 3). He shared his expertise and greatly influenced Grajo's Farm to embrace the concept of vertical hive design. He later migrated to Australia and upgraded his design to suit Australian stingless bee species and is currently the business owner of Nativo Bees (Venturieri, 2008).

Figure 1. Traditional method of stingless beekeeping using coconut shells.
Source: The Colony by Grajo's Farm

Figure 2. Traditional method of stingless beekeeping using coconut shells.
Source: The Colony by Grajo's Farm

Figure 3. Dr. Giorgio Venturieri
Source: Dr. Giorgio Venturieri

VERTICAL HIVE DESIGN

The vertical hive was designed by Angolan Professor Virgílio de Portugal (de Araújo Filho, 1957)and (Goff, 1976) (Fig. 4). The concept of this square hive box model is based on the natural hive pattern on hollow trees. The chambers are limited into two: a lower chamber for brood combs and food storage on the top, usually called "melgueiras". On the bottom left corner is the inlet of air and on the left top is the outlet. The entrance is found on the right corner passing through a tube-like structure and leading to the brood chamber also called the "gallery" (1).

Figure 4. The hive design by Portugal - Araujo showing a brood chamber on the base and a food storage on top.
Source: Com.tec212.pdf (embrapa.br)

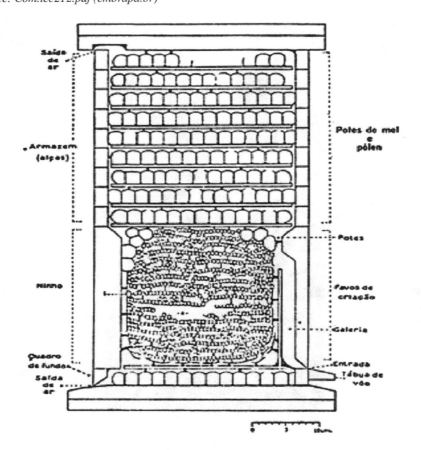

Figure 5. Portugal Araujo hive design modified by Dr. Giorgio Venturieri
Source: https://www.researchgate.net/publication/312351183

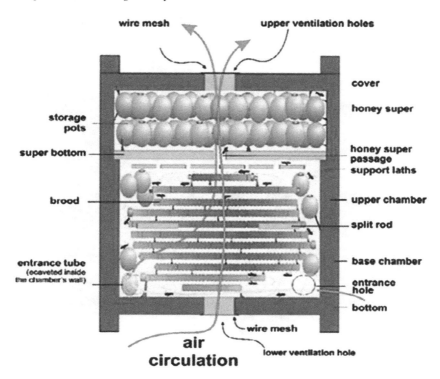

Dr. Venturieri modified the vertical hive design of Prof. Portugal Araujo by altering the top lid and base ventilation to allow air to flow in the hive for heat and moisture regulation (Fig. 5). He called this Modified Portugal Araujo Hive Design to give credit to the original designer. In addition, he also took account of other modifications of several researchers such as Alvarez et al. (2018) and de Oliveira (2012).

According to Venturieri (2004), the natural shape of the brood is rounded (Fig. 6). However, a square hive box helps the bees regulate the temperature in the middle of the brood since the incubation of the immatures needs to be constant around 28-30°C. Furthermore, the hive dimension is based on the diameter of the brood, adding two centimeters on all sides (Fig. 6). Ideally, one inch and 1.33 cm thickness are recommended for tropical and temperate countries, respectively.

Wood panels/frames are uniform in dimension to minimize offcuts and facilitate easy cutting. The panels are joined using the technique called butt joints in which ends of each wood panel/frame are joined and nailed together to form a square box (Fig.)6. Since this joint is weak, the separators increase

Figure 6. a-b) Tetragonula biroi combs with rounder shape (photos above); c) yellow arrows represent 2 cm allotted space that serves as passage upwards so bees can store food on top, d) blue arrows show the butt joint technique in which the panels overlap and nailed together (photos below).
Photo: The Colony by Grajo's Farm

its strength by providing support and stability to the chamber/ hive box using marine plywood with a thickness of 0.25 inch and rabbeted in the bottom panel/ frame of the chamber (Fig. 7).

When the Modified Portugal Araujo Hive Design by Contrera et al., (2011) was first adopted for Tetragonula biroi by Grajos's Farm, wood panel/ frame measures 8.0 inches long, 0.75 inches wide, and 3.0 inches tall while the assemble chamber measures 8.75 inches long, 8.75 inches wide, and 3.0 inches tall. This is based on the average brood diameter of the five experimental hives. This design is currently updated to meet the recommended thickness and height of the chamber. The outer dimension was not changed to ensure that it still fits the old hive boxes. The upgraded version is of the hive chamber measures 8.75 in long x 8.75 in wide and 4 in height

Figure 7. The first model of Dr. Venturieri's Modified Portugal Araujo hive design adopted for Tetragonula biroi by the Grajos' Farm.
(Photo: The Colony by Grajo's Farm)

Figure 8. The inner and outer dimension of the adopted modified vertical hive design for Tetragonula biroi.
(Photo: The Colony by Grajo's Farm)

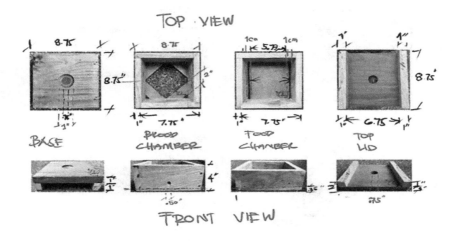

(individual wood panel/ frame is 7.75 in long x 1.0 in wide x 4.0 in tall) provides a liquid volume of 2.83 liters per chamber (Fig.).

The Philippine cedar (Toona kalantas), known as light, durable, and weathering-resistant, has been the Grajo's Farm wood choice ever since (Fig. 9). Unfortunately, due to scarcity of supply and high cost, they shifted to using Palochina, a cheap softwood coming from recyclable wooden pallets (Fig. 9). After cutting Palochina

Figure 9. The inner and outer dimension of the adopted modified vertical hive design for Tetragonula biroi.
(Photo: The Colony by Grajo's Farm)

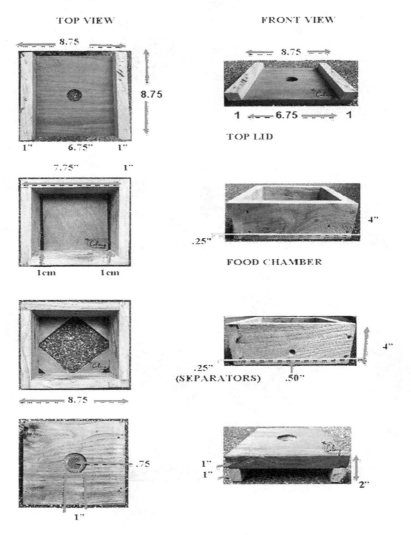

Figure 10. a) The Modified Portugal Araujo hive design by Dr. Venturieri adopted for Tetragonula biroi b) upgraded version of the modified hive design for pollen and honey production.
Photo: The Colony by Grajo's Farm

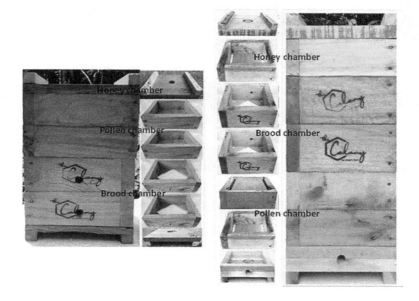

into desired dimensions, it is treated with a brine solution to prevent woodborer infestation, air-dried, and later assembled into hive boxes.

Hive Parts

The hive design allows the colony to grow upwards, passing through the brood chambers/ modules going up the food chambers. Therefore, harvesting is done on top. However, the newly adopted design is introduced primarily for the production of pollen and honey. Food chambers are split into two for honey production (top part) and pollen production (bottom part). Modified base, entranceless brood chamber, and double separators are also created (Fig. 9).

a. Base

The base has an orifice that enables the flow of air within the hive (Fig. 10). A thin metal wire mesh is installed to prevent bees from making secondary entrances (1)(6). Ideally, the base should be detachable for easy cleaning and draining. The modified base is another Brazilian concept (Fig. 10) containing a slightly angled entrance (Fig.

Figure 11. a) original base, b) modified base containing hive entrance with a slightly angled entrance and c) close up photo of the slightly angled hive entrance.
Photo: The Colony by Grajo's Farm

10). It protects the direct entry of water into the hive during the monsoon season and slows down the entry of pests.

b. Brood Chamber

The brood chamber houses the brood combs and a couple of food pot reserves for brood development. The rhombus shape serves as a separator and, at the same time, helps to support the brood and food pots that are essential in the success of colony duplication (Fig. 11) (1, 5, 6). The entrance of the first adopted design is located on the center front of the brood chamber (Fig. 9). A separator is made of ¼ thick marine plywood, which is rebated into the panels/ frame of the chamber (Fig. 11) to ensure that there are no gaps or crevices between the chambers. Hence, pests are prevented from laying eggs in the gaps.

Both circle and rhombus shapes have the same function, but the rhombus type is more cost-effective (Fig. 11). However, the modified brood chamber design for pollen and honey production is entranceless and relocated into the new base design (Fig. 10) to simulate their habit of storing food, particularly pollen, underneath the brood for brood development.

c. Food Chamber

Pollen and honey are deposited in this chamber. A separator or excluder is installed to limit the growth of the brood. The two spaces on both sides allow bees to access the reserved space for food accumulation (1,5). The first adopted design is composed only of a permanent separator at the bottom (Fig. 12). For pollen and honey production, double separators are used. The bottom is fixed while the top separator is detachable, to strategically interchange food chambers and facilitate harvesting (Figure 12).

Figure 12. a- b) Top, rear view and rebated rhombus separator on the bottom frame brood chamber, c-d) rounded type separator rebated also in the brood chamber.

Figure 13. a) adopted food chamber design for Tetragonula biroi, b) double separator that can be placed either on top or bottom of the brood.

d. Top Lid

The top lid contains an orifice measuring ¾ inches located at the center to serve as ventilation. A metal wire mesh is installed to cover the orifice so bees will not establish another entrance (Fig. 13) (1). The bees construct an interesting ventilation structure when the orifice is not covered with a mesh screen, as shown in Fig. 13.

Figure 14. a) Top lid with metal screen and b) ventilation hive with mesh screen
Photo: The Colony by Grajo's Farm

Advantages

a. Productivity

To accelerate the brood production, hence, fast duplication of colonies, the hive dimension was based on the diameter of the brood.

b. Climate Resiliency

It contains ventilation holes that help air circulation. In addition, the thick wood used helps regulate temperature and moisture in the hive. Since the hives are individually roofed, they are well protected from rain and sunlight. The hives are also strapped on stands with a maximum capacity of ten hives per stand. After the typhoon, fallen hives are just reinstalled and re-strapped back to their original position.

c. Management

The hive box comprises functional parts such as base, brood, and food chambers that facilitate easy manipulation during duplication and harvesting. The brood chamber contains a separator that supports and protects the brood combs, while the food chamber contains a separator or excluder that prevents the brood growth on food chambers. In addition, the base and top lid are removable for easy cleaning and inspection. The 'minimal disturbance method' is a natural hive duplication propagation technique that is non-invasive, resulting in fast colony splitting and less worker mortality during the harvesting and duplication process. This method prevents physical contact with brood combs and food pots during duplication and harvesting, resulting in non-contamination.

d. Pest and Disease Resistance

The food and brood chambers are flushed to each other, not creating gaps or crevices to prevent pest invasion and keep the colony strong. Due to the very quick duplication method, pest incidence is reduced. A strong colony is not prone to infestation. Usually, natural enemies of the stingless bees are propolized/ mummified by the workers inside the hive.

e. Durability

The Toona kalantas is the best wood choice as it is weathering-proof. Thicker wood would last longer than thin wood as it is prone to chipping through time. Palochina wood needs treatment to improve the lifespan and resistance to wood-boring insects.

Modifications in the Hive Design

Modifications were made in the design to address the various challenges met in stingless beekeeping, such as failure in feral to hive box colony transfer, colony mortality, prolonged rain, and pest infestation. The changes made in the design are the following:

1. The entrance hole is slightly angled to prevent direct entry of water into the chamber and slow down pests' entry.
2. The separators were rebated in the mainframe of the chambers to flush them with the other chambers, thereby minimizing the chance of pests laying eggs on the gaps and crevices between chambers.
3. The panels/frames of the chambers are cut uniformly and nailed together using the butt joint technique to minimize offcuts and save labor, time, and effort instead of making different sizes of panel/frames for the chambers. Doing this makes assembling easy, too.
4. The chambers for pollen and honey production are re-arranged. The food chamber on top of the hive containing pollen is relocated to the bottom near the brood, functioning as a reserve chamber that contains fermented bee bread. In addition, a harvestable pollen chamber is introduced underneath the reserve pollen chamber for bee bread production.

5.   A new base comprising the entrance allows easy manipulation and gets utilized for food and brood chamber. The space within the base will provide a draining system and keep the bottom ventilation unblocked when using the food chamber. Also, insect traps can be placed in the base as a preventative measure.

MANAGEMENT PRACTICES

The establishment of meliponary at Grajo's Farm started as a hobby and bloomed as the years went by. The passion for stingless bees grew even more, when the number of colonies started to increase. The progress may have been slow but rewarding as they learned how to manage colonies well for the past 11 years. Below are the management practices being employed on the farm.

Feral Colony Hunting

Through the years, they teamed up with bee hunters. They taught them proper feral colony hunting techniques to minimize mortality during the extraction, thereby preventing pest infestation. The hunted colonies are first placed in sickbay under observation for a couple of weeks near the bee hunter's backyard. The small/ weak colonies are separated from larger and strong colonies. Later, the colonies are brought to the farm, isolated again from the existing stingless bee stands, and classified bases on colony strength. The colonies are allowed to undergo a recovery period for at least one to two months.

Installation of Duplication Brood Chambers on Feral Colonies

After the recovery period of 1-2 months, stored pollen and honey are left untouched and become food reserves. This practice helps accelerate the production of brood upwards to the new chamber provided. The brood combs of the feral colony are exposed, and a permanent chamber/ feral extension is installed (Fig. 13a), usually.

Colony Division Colonies Using Feral

Rather than destroying the colony by transferring the brood cells forcefully that causes stress to the bees, an easy and less destructive method of colony division unharmful for the bees is practiced at Grajo's farm. This method was adopted for feral colonies after introducing the modified vertical (Kiros & Tsegay, 2017) (Figure 14). This natural hive duplication technique efficiently produces 2-3 new colonies per year and preserves the colony for at least 5-8 years. Using the hive box design,

Figure 15. a) installation of brood chamber extension on a feral colony, b) additional duplication brood chamber is added on top of extension chamber, c) food chamber is added on top of filled duplication chamber, d) once the duplication and food chamber are full this can be splitted/ duplicated into a new colony.
Illustrations by: Leo Grajo

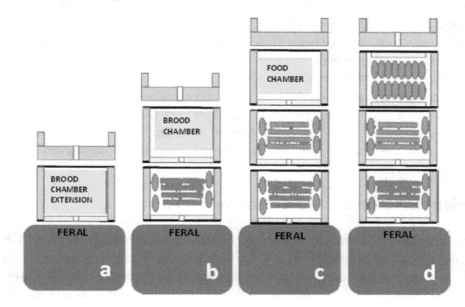

Figure 16. a) the duplication chamber and food chamber are removed using the minimal disturbance method leaving the extension brood chamber in the feral colony. b) duplication brood chamber is added on top of the feral extension chamber and gradually adding the food chamber again.
Photo: The Colony by Grajo's Farm

colony manipulation using chambers containing specific and functional separators or excluders became easy.

Once full, the duplication chamber and food chamber are removed to form a new colony. The newly duplicated brood chamber is mounted on the base, an empty brood chamber is added in the middle of the food and brood chambers for brood comb expansion and the food chamber is placed on top of the empty brood chamber to serve as a food reserve (Fig. 15). After removing the new colony, a new empty duplication brood chamber is again added on top of the feral's extension/permanent chamber, and the colony production cycle starts again (Fig. 15 and 15).

After years of colony production, the brood chamber within the feral colony goes up and is replaced with pollen. When this happens, the permanent / extension chamber is removed from the feral colony with the top brood and food chambers and becomes the mother colony. The pollen, honey, and propolis are harvested from this old feral colony.

Colony Division Using Modified Vertical Hive Box

This natural duplication technique is very common in Brazil and was proposed by (E. M. de Oliveira & Oliveira, 2018), calling it the "minimal disturbance method" (Fig. 16). It results in a quick duplication, faster recovery, reduced chance of infestation (5).

In addition, Dollin & Dollin (1997) publication on natural duplication techniques in feral colonies became the inspiration of Grajo's Farm. The candidate colony must be strong for a higher chance of survival. Since the first adopted design for hive

Figure 17. a) new duplicated colony, b) three-layered brood chamber and a food chamber, c) the 2 bottom brood chamber that forms a mother colony.
Photo: The Colony by Grajo's Farm

boxes only measures three inches in height, Grajo's Farm ensured at least three layers of the brood chamber and at least one food chamber before duplication. As shown in Fig. 16a, the left side contains an advancing front and an empty brood chamber (with the yellow arrow) placed in the middle. This positioning is altered based on the original duplication method shown in Fig. 15 by installing an empty brood chamber on the middle for colony expansion and a reserve food chamber on top (Fig. 16a). It is done to ensure that the new colony has enough food to survive and is placed back where the mother colony is situated to receive the returning foragers. The right side contains the combs where mature ones at the near-hatching stage are located to recover effortlessly, becoming the mother colony (Fig. 16b and 16c). Either empty food or brood chamber can be placed on top of the mother colony. The mother colony is moved away at least ten meters from the original position.

Colony Inspection

The newly duplicated colony is inspected after 20 days or once a month. The colony's health is monitored to determine the presence of a queen and pests. The results of the inspection and monitoring of colonies are essential in deciding the best intervention needed, including possible duplication and harvesting.

Harvesting and Processing

Harvesting is done at the same time as colony duplication. Since the primary goal of Grajo's Farm is colony multiplication, only the excess honey and pollen are harvested, otherwise known as selective harvesting, especially in flow season.

a. Honey Harvesting

The honey is extracted on-site using clean, sterilized tools and stored in food containers. It is pre-checked and segregated from pure and with pollen mixture. Honey containing minimal pollen pots is removed instantly to avoid contact with pollen to minimize fermentation and ensure the purity of the honey. The honey pots are brought to the processing center, where they will be pressed, strained, bottled, and labeled (Fig. 18). After bottling, the pure (premium) honey is placed into a refrigerator to stop fermentation. According to Contrera et al., (2011), the honey contaminated with pollen can either be placed in the refrigerator to settle and can be scope later or let it matures with pollen, also it can go pasteurization. Honey is pasteurized at 62 Celsius for 10 minutes to stop the fermentation and bottled using a double boiler. The pasteurized honey is placed at room temperature without fermentation (6). Later, labeling is done to improve the marketability of the product.

Figure 18. Pasteurized (left) and left) andaw (right) stingless bee honey.
Photo: The Colony by Grajo's Farm

Figure 19. Granulated (upper left photo) and solid form (upper right photo) of bee breaof beeged in bottles (lower photos)
Photo: The Colony by Grajo's Farm

b. Pollen Harvesting

Like honey, dry and moist bee bread is separated from and placed into a cleaned and sterilized container. The bee bread is brought again into the processing center for processing. The bee bread is peeled off from the cerumen pots and placed in a food dehydrator to reduce the moisture content of the bee bread for 24 hours. The dehydrated bee bread is separated. The intact bee bread ball is also bottled directly.

Figure 20. Associated pests (insects and frog) found within and outside the colony feeding on stingless bees. a) Sap beetle b) Black soldier flies c) common cricket, d) Assassin bugs, e) Toad.
Photo: The Colony by Grajo's Farm

The damaged/ broken bee bread will undergo another process using a grinding machine to produce granulated bee bread (Fig. 19).

Pest and Disease Management

In the early years of the establishment, pests on the farm (Fig. 20) were encountered but they managed to eliminate them through the years. Prevention is the key in combating these pests by keeping the colony strong, avoiding over- density, selective duplication/splitting, using good hive design (no gaps or crevices where pests can lay eggs), installing traps, and quick duplication process.

Colony Packaging

The starter colony is made up of two brood chambers and one food chamber containing less food. The entrance is sealed and packed at night using a bag before the pickup day.

Colony Transport and Marketing

Before transporting hived stingless bee colonies, a Veterinary Health Certificate (VHC) is secured from the Bureau of Animal Industry (BAI). A BAI representative conducts inspection and collection of samples for clinical laboratory examination. The result is forwarded to the Provincial Veterinary Office to issue the local shipping permit with a 5-day validity from the date of issuance.

CONCLUSION

In conclusion, this study has observed that some of the stingless bee species on the farm produce small broods in terms of diameter. Proper naming and identification of stingless bees are then needed to design an appropriate hive dimension fit for a particular species. Currently, chemical-treated marine plywood is used as a separator. There is a need to look for other woods that are chemical-free and safe for the bees. Coating of hive boxes is also necessary to extend its life and prevent hive boxes from weathering. The modified hive design for pollen and honey production in two food chambers is purposely placed underneath the brood chambers for a food reserve. The harvestable pollen chamber seems promising and needs further trial.

LIMITATIONS AND CHALLENGES

Lack of Technical Knowledge and Beekeeping Experience

During the early stage of the establishment, high colony mortality was associated with the lack of technical knowledge on the proper transfer of brood combs to the hive boxes. The feral colonies were not allowed to recover from handling disturbance. Boxing feral colonies even if they have been damaged during the extraction from the wild. Damaged feral colonies are the primary target of pests. Some of these colonies have not been properly handled during the extraction from the wild resulting in broken pots leaking towards the brood, drowning the brood, or accidentally falling to the ground. It is advisable to use sanitized tools when splitting colonies and harvesting hive products. Also, avoid throwing excess pollen pots anywhere in the meliponary site for it can attract and serve as a haven for egg-laying insect pests.

Several Meliponini houses were established during the early years to house hives more than the area could carry, leading to the loss of colonies due to robbing and shortage of foraging areas. Premium timber demands a high price due to its scarcity. Bicol Region is frequently visited by typhoons and is geographically located on the typhoon route in the Philippines. In addition, the long rainy season makes it hard for the bees to forage.

Lessons Learned

Adoption of the natural hive duplication technique on feral colonies and the minimal disturbance method should be done as it is very efficient in the colony, pollen, and honey production. A permanent extension chamber should be introduced to a feral colony to facilitate the easy duplication of the new colony without damaging it during the duplication. A two-week initial recovery period should be imposed when buying feral colonies by sealing the opening of the feral colonies on the bee hunter's place before bringing it to the farm to minimize pest infestation. New feral colonies should be isolated from the existing colonies on the farm and classified according to their size and strength. Another one- to two-month final recovery period should be done before the installation of the permanent chamber. The permanent extension chamber installation on the exposed provision should be placed on top to ensure that it is stable and batumen is sealed around the permanent brood chamber and then taped.

The beekeeper should consider the carrying capacity of the farm to avoid colony mortality due to over-density. The distribution of colonies in more than one meliponary is necessary to minimize losses caused by robbing and invading swarms. When the feral colonies are installed with duplication chambers, it can provide 2-3 colonies per year and 5-8 years of continuous production of new colonies. The rate of production depends on feral colony size, food sources, and climatic conditions. Swarming of colonies is likewise minimized because of fast colony reproduction. On the other hand, the modified vertical hive boxes can produce 2-3 colonies per year, depending on colony strength.

Due to the scarcity and high price of premium wood, one may look for an alternative such as palochina timber. Palochina coming from recycled wooden pallets or crates is the best alternative. But these are prone to wood borer; thus, soaking it in a brine solution rather than air-drying it is necessary. For typhoon-prone areas, build durable bee stands and place them around the periphery of the farm to disperse the colonies for the stingless bees to maximize the foraging area. With durable stands, colonies are secured, and bagging and packing of colonies before the typhoon is no longer necessary. In addition, provide a roof to each colony to protect it from rain and sun.

ACKNOWLEDGMENT

The author expresses his sincerest thanks to his sisters, Leizel Grajo, Kathleen Grajo, and Yul Manel Grajo, for the support and help in managing the farm; Jairo Come and Christopher Osma, for being reliable beekeepers on the farm; Dr. Cleofas R. Cervancia, for her encouragement to engage in stingless beekeeping; Dr. Amelia R. Nicolas, for her guidance in writing and editing the manuscript; Dr. Giorgio Venturieri, for being his mentor who unselfishly shares his expertise not just to him but to the rest of the world in terms of stingless beekeeping; and lastly, to his loving wife, Therese, and son, Liam, for being his source of inspiration and solid support.

REFERENCES

Abante, C. G. R. (2020). Mayon volcano cultural heritage as a source of place identity in Guinobatan, Albay, Philippines. *Bicol University R&D Journal*, *23*(1), 1–14.

Alvarez, L. J., Reynaldi, F. J., Ramello, P. J., Garcia, M. L. G., Sguazza, G. H., Abrahamovich, A. H., & Lucia, M. (2018). Detection of honey bee viruses in Argentinian stingless bees (Hymenoptera: Apidae). *Insectes Sociaux*, *65*(1), 191–197. doi:10.100700040-017-0587-2

Baroga-Barbecho, J. B., & Cervancia, C. R. (2019). *Pest of Philippine stingless bees*. Philippine Entomologist.

Belina-Aldemita, M. D., Opper, C., Schreiner, M., & D'Amico, S. (2019). Nutritional composition of pot-pollen produced by stingless bees (Tetragonula biroi Friese) from the Philippines. *Journal of Food Composition and Analysis*, *82*, 103215. doi:10.1016/j.jfca.2019.04.003

Cervancia, C. R. (2018). Management and conservation of Philippine bees. In *Asian beekeeping in the 21st Century* (pp. 307–321). Springer. doi:10.1007/978-981-10-8222-1_14

Contrera, F. A. L., Menezes, C., & Venturieri, G. C. (2011). *New horizons on stingless beekeeping (Apidae, Meliponini)*. Academic Press.

de Araújo Filho, J. R. (1957). A cultura da banana no Brasil. *Boletim Paulista de Geografia*, *27*, 27–54.

de Oliveira, E. M., & Oliveira, F. L. C. (2018). Forecasting mid-long term electric energy consumption through bagging ARIMA and exponential smoothing methods. *Energy*, *144*, 776–788. doi:10.1016/j.energy.2017.12.049

Dollin, A. E., Dollin, L. J., & Sakagami, S. F. (1997). Australian stingless bees of the genus Trigona (Hymenoptera: Apidae). *Invertebrate Systematics*, *11*(6), 861–896. doi:10.1071/IT96020

Goff, R. C. (1976). Vertical structure of thunderstorm outflows. *Monthly Weather Review*, *104*(11), 1429–1440. doi:10.1175/1520-0493(1976)104<1429:VSOTO> 2.0.CO;2

Kiros, W., & Tsegay, T. (2017). Honey-bee production practices and hive technology preferences in Jimma and Illubabor Zone of Oromiya Regional State, Ethiopia. *Agriculture and Environment*, *9*(1), 31–43.

Oliveira, L. K. de. (2012). *Energia como Recurso de Poder na Política Internacional: geopolítica, estratégia e o papel do Centro de Decisão Energética*. Academic Press.

Venturieri, G. C. (2004). *Criação de abelhas indígenas sem ferrão*. Embrapa Amazônia Oriental.

Venturieri, G. C. (2008). *Caixa para a criação de uruçu-amarela Melipona flavolineata Friese, 1900. In Embrapa Amazônia Oriental-Comunicado Técnico*. INFOTECA-E.

Chapter 6
Comparison of Total Soluble Protein Content and SDS–PAGE Pattern Between Four Different Types of Honey

Fisal Haji Ahmad
Universiti Malaysia Terengganu, Malaysia

Mohd Amiruddin Abdul Wahab
Universiti Malaysia Terengganu, Malaysia

Tuan Zainazor Tuan Chilek
Universiti Malaysia Terengganu, Malaysia

Amir Izzwan Zamri
Universiti Malaysia Terengganu, Malaysia

Shamsul Bahri Abd Razak
Universiti Malaysia Terengganu, Malaysia

Azril Dino Abd Malik
Naluri Pantas Sdn. Bhd, Malaysia

ABSTRACT

Generally, there are two types of beekeeping: the Apini tribe and the Meliponini tribe. Both tribes produce honey and have a good demand due to their health benefit properties. Considering the influence of diverse factors on honey composition and the lack of studies, establishing quality standards for stingless bee honey (Meliponini tribe) is still challenging and need to do to protect the consumer. In this sense, this

DOI: 10.4018/978-1-6684-6265-2.ch006

study aimed to determine the total soluble protein content and compare the SDS-PAGE profile between two species of Apini tribe and two species of Meliponini tribe. Protein concentrations in honey samples were varied and resulted in a micro component in honey. SDS-PAGE profile for Meliponini tribe showed more number of protein bands compared to protein from Apini tribe. The unique protein bands that appeared in the Meliponini tribe may have potential as a biomarker to justify the authenticity and quality of that honey, which is known as Unique Kelulut Factor (UKF).

INTRODUCTION

Honey is a natural sweetener that is consumed worldwide. Malaysia has various types of honey that are already commercial in the global market. There are two types of honey which blossom honey and honeydew honey. The former is produced from the nectar of flowers, while the latter is produced from secretions of the living parts of plants other than flowers or is a product of bees' excretions (Azevedo et al., 2017). Honeybees (Apini tribe) and stingless bees (Meliponini tribe) are two common main honey producers. Honeybees, also known as Apis mellifera bees, are normally bigger and sting, while stingless bees such as Heterotrigona itama are smaller in size and without a sting. As Naila et al. (2018) mentioned, Apis mellifera bee honey is sweet, while stingless bee honey is a mixture of sweet and sour tastes.

Generally, honey is a viscous solution containing various molecules, including fructose and glucose (80-85%); water (15-17%); ash (0.2%); proteins and amino acids (0.1-0.5%) and trace amounts of enzymes, vitamins, and other substances, such as phenolic compounds (Rao et al., 2016). According to Alimentarius (2001), honey is the natural sweet substance produced by honeybees from the nectar of plants, from secretions of living parts of plants, or from excretions of plant-sucking insects. This official definition is restricted to the honey produced by A. mellifera. It may not apply to stingless bee honey, better known as pot honey, which is very popular for its distinct sweetness mixed with an acidic taste, and fluid texture (Ávila et al., 2019). However, the honey composition varies depending on the plants from which the bee consumes nectar. For this reason, the classification and evaluation of honey have always been a challenge for chemical analysis, especially when honey adulteration is increasing (Ramón-Sierra et al., 2015). Among these two types of bees that produce honey, the demand for stingless bees compared to Apis bees is increasing nowadays because of their medicinal value, which can act as an antimicrobial, anti-cancer, anti-inflammatory and wound healing (Queiroz-Junior et al., 2015).

Despite the traditional uses of honey as a therapeutic agent, honey has recently been recognized in the development of modern medicine for its valuable nutritional quality (Rao et al., 2016). The common therapeutic properties of most honeys are mostly due to their floral origins. In addition, to its high nutritional value, honey also exhibits significant health benefits such as wound healing, antimicrobial, antioxidant, anti-inflammatory, antidiabetic, and anticancer effects. For example, the benefits of Manuka honey, European bee honey, have been recognized internationally (Zulkhairi Amin et al., 2018). Although there is only a small amount of protein and its byproducts (enzymes, bioactive peptides, and amino acids) available, they are also responsible for the health benefits of honey besides the major components such as flavonoids, phenolic acids, and ascorbic acid.

The use of proteomics approaches is a powerful tool in food science regarding process optimization and monitoring, quality, traceability, safety, and nutritional assessment (Pedreschi et al., 2010). Proteins, together with peptides, are one of the major groups of food components, and they are found in many different organisms of both plant and animal origin. The analysis of the complete proteome of foods can characterize foods with unique quality, food produced by new technologies or food originating from specific geographical areas (authenticity). Proteomics has also been used in market surveillance of genetically modified foods. Another new challenge for proteomics has recently been recognized in studying molecular interactions and differences in food proteomes relevant to nutrition (Corradini et al., 2010). As in the case of food analysis, proteomics techniques present several advantages and can provide useful information to identify proteins in food matrix, to study protein–protein interactions in raw and processed foods, as well as interactions between proteins and other food components, or to identify covalently bound constituents that may be produced during processing (Ibáñez et al., 2012).

A typical proteomics workflow consists of sample preparation/protein extraction, protein separation, staining/quantification and image analysis, mass spectrometry and data analysis and interpretation (Carpentier et al., 2005). Due to the wide variety of protein sample types and sources, an appropriate sample preparation method is required in which the samples are treated before their analysis. Sample preparation profoundly affects the outcome of protein and peptide separation and their subsequent analysis. These procedures need to be compatible with the following analysis by two-dimensional electrophoresis (2D-PAGE) and/or liquid chromatography-tandem mass spectrometry (LC-MS/MS) (Carpentier et al., 2005). Therefore, this study selected method extraction to be compatible with proteomics workflow.

Honey protein originates from the bees' hypopharyngeal glands and salivary glands of bees and from the enzymatic reaction of pollen and saliva (Naila et al., 2018). Therefore, protein in honey might contribute to specific biomarkers and can be used for authentication purposes. As it is known, the protein content in honey is low and high with non-protein components; therefore, it is necessary to have a good protein extraction method that can remove all the unwanted components. Few tests have been used to differentiate adulterated and pure honey using the honey protein (Jamnik et al., 2012). In a previous study by Šimúth et al. (2004), the major protein of honey, apalbumin-1 was proposed as a marker for immunochemical testing to reveal adulterants in honey. Besides, honey enzymes may be used to develop a biosensor that could detect the freshness of honey. Nonetheless, active enzymes in the honey are necessary to determine the authenticity of honey because they may become inactive during storage or when honey is heated. Furthermore, proline is usually the major amino acids found in honey which is comparable to up to 90% of the total free amino acid content and has been suggested as a ripeness indicator of honey (Seraglio et al., 2019).

The determination and identification of proteins in honey by SDS–PAGE and proteomics have been considered less as an index of quality control of honey. There is little data concerning markers of entomological origin using this method. Since the soluble protein content in honey is low and to the best of our knowledge, the protein in honey studies are not very intensively done. In this sense, this study aimed to determine and compare the total soluble protein content, SDS-PAGE pattern and amino acid profile between 4 types of honey available in the market. Two samples from Apis tribe were Apis mellifera (Manuka honey) and Apis dorsata (Tualang honey), and two from the Meliponini tribe were Geniotrigona thoracica and Heterotrigona itama. Thus, this research contributes to the literature as fundamental data regarding protein from different honey that can reflect the quality and authenticity of the honey for future reference.

MATERIALS AND METHODS

Sample Collection

Four samples of honey were purchased from the selected manufacturer in August 2020, namely Apis mellifera (Manuka honey), Apis dorsata (Tualang honey), Geniotrigona thoracica and Heterotrigona itama (Kelulut honey). The honeys were stored at refrigeration temperature throughout the experimental period. Visual inspection of the samples did not show any crystal formation during storage.

Extraction of Protein

The extraction method was prepared according to Bocian et al. (2019) with a slight modification. 2g of honey was diluted with 2g of double-deionized distilled water 1:1 (w/w). After the honey solutions obtained, the solutions were transferred into 50ml centrifuged tubes. Using weight as the measurement for honey is more efficient since it has different viscosity and density among each type of sample. Then, the samples were mixed with 3mL of ice-cold Tris-buffered (pH 8.0) phenol solution (1:1) v/v. The addition of phenol solution must be takes place in a fume hood as the phenol is a highly hazardous chemical. The solutions were vortexed for 15 minutes to ensure that all the protein was soluble in phenol. Samples were then centrifuged at 10,000 rpm for 10 minutes. The temperature during centrifuge was set at 4°C to avoid heat denaturation caused by high spinning speed. Samples were gently removed from the centrifuge to prevent the solution from mixing again.

The upper phase was taken and transferred into a new 50mL centrifuge tube. The skill of pipetting the upper phase solution was crucial in this transfer process. All the liquid from the upper phase must be taken carefully without sucking the layer between the phases. The upper phase of liquid in all tubes was mixed with ice-cold 0.1M ammonium acetate in methanol 5 times the volume of the liquid obtained. The solutions could precipitate overnight at 4°C. After the overnight precipitation, the solutions were centrifuged at 12,000 rpm for 30 minutes. The supernatant was discarded, the remaining pellets only. The pellets obtained were washed using 2mL of ice-cold acetone dithiothreitol (DTT). The samples were incubated at 4°C for 1 hour. Samples were centrifuged again at 12,000 rpm for 20 minutes. Supernatant were discarded carefully in fume hood. Pellets were set to air dried in a fume hood. Then, the protein pellets were stored at -80°C deep freezer before resolubilize again. The air-dried pellets were resuspended in 350 µL solution containing 7 M urea, 2 M thiourea, 4% CHAPS, and 1% DTT. The sample mixture was gently vortexed every 15 mins for 1 hour at room temperature. When necessary, they were cooled in a chiller to prevent sample heating. The supernatant was collected after centrifugation at 10,000 ×g for 30 min at 4oC and was stored at -80oC in airtight containers until further analysis. Estimation of total soluble protein concentration was done before storing.

Estimation of Total Soluble Protein Concentration and Protein Yield

Total soluble protein in the seaweed samples was determined using the Bradford assay (Bradford, 1976) on 96-well microtiter plates with some modifications referring to the Bio-Rad Protein Assay Manual. Protein concentration was calculated according to the protein standard curve of bovine serum albumin (BSA). The Bradford reagent (BIORAD, Hercules, CA, USA) was diluted five times with distilled water before use. About 200 µl of the diluted Bradford reagent was added to all standards (BSA) and protein extracts (10 µl), mixed and incubated for 5 min in the dark at room temperature (26 oC ±1). The absorbance at 595 nm was measured after 5 min reaction in microtiter plates according to the manufacturer instructions using a Thermo Scientific microplate reader (Multiskan Ascent V1.25 Plate Reader with Ascent Software version 2.6: Thermo Electron Corporation, Vantaa, Finland).

A standard curve was plotted in Microsoft Excel, and the slope equation was used to determine the protein concentration in the samples. Using the calibration curve, the protein concentration of each sample was determined. The absorbance value at 595 nm is directly proportional to the amount of solubilized protein in each sample.

Figure 1. A calibration graph of BSA standards for protein quantification.

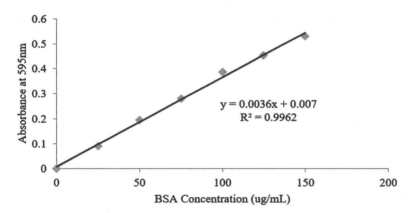

Figure 1 presents a calibration graph obtained using a series of BSA standards. BSA is commonly used as a protein standard because it is inexpensive and readily available in pure form (Kruger, 2002). The ubiquity of BSA as a protein standard allows this study's results to be compared directly to those of many previous studies. The protein concentration was expressed as microgram per microliter (µg/µL). Protein yield was then calculated and stated as microgram per gram of honey weight (µg/g honey) with three replicates and presented as a mean ± standard deviation.

SDS-Polyacrylamide Gel Electrophoresis

Protein-denaturing sodium dodecyl sulfate-polyacrylamide gel electrophoresis (SDS-PAGE) was performed according to (Laemmli, 1970) with some modifications using mini gels with 8.3 x 7.3 cm dimension and 1 mm of thickness. SDS-PAGE was performed in a Mini-Protean Tetra Cell Electrophoresis apparatus (BioRad, USA) using resolving and stacking polyacrylamide gels with the compositions of 12.5% acrylamide in the resolving gel. The stacking gel (with a large pore size) concentrates all of the proteins (the large ones can catch up with the small ones) on top of the resolving gel. After entering the resolving gel (which has a smaller pore size), the proteins are separated according to relative molecular size.

The protein extracts were mixed with an equal volume (1:1) of Laemmli sampling buffer containing 62.5 mM Tris-HCl (pH 6.8), 25% v/v glycerol, 2% w/v SDS, 5% v/v β-mercaptoethanol and 0.01% w/v bromophenol blue. The samples were mixed in a microcentrifuge tube and heated in boiling water for 5 min. The ß-mercaptoethanol in the sampling buffer reduces the protein disulfide bonds, and the SDS denature the proteins. The sampling buffer contains glycerol to increase the density so that when the sample is loaded it sinks to the bottom of the well. Bromophenol blue dye is in the sampling buffer to monitor the electrophoresis process.

Following the polymerization of the stacking gel, the gel sandwiches were inserted into the running tank. Approximately 300 mL of electrophoresis running buffer was poured into the lower buffer chamber, and additional electrophoresis running buffer was added to the inner cooling core, which served as the upper buffering chamber. The plastic comb was removed from the gels, and a selected amount (to be fixed in all samples) of each treated protein sample was loaded into the individual wells formed in the stacking gel. One well in the stacking gel was loaded with 5 μL of a molecular weight marker standard (Bio-Rad, USA) (10 kDa to 250 kDa) in order to determine the molecular weight of protein fractions present in the extracts. The electrophoresis process only can be run after all samples have been loaded.

The Mini-Protean Tetra Cell Electrophoresis apparatus (BioRad, USA) was connected to an Electrophoresis power supply Power Pac Basic (Bio-Rad, USA) and the electrophoresis was carried out at a constant voltage of 180 V for approximately 1 hour by which time the bromophenol blue tracking dye had migrated to the bottom of the gel. Following electrophoresis, the gels were removed from the gel sandwiches and placed into the gel staining container and ready to be stained.

Coomassie Blue Staining

Polyacrylamide gels were Coomassie blue (CBB) stained as described by Laemmli (1970) with some modification. Briefly, the gel was incubated in a staining solution containing 40% v/v methanol, 10% v/v acetic acid and 0.1% w/v Coomassie Brilliant Blue R250 with gentle shaking overnight. After incubation, the dye solution was removed, and the gel was rinsed with ultra-pure water 1-2 times to remove the dye.

After that, a destaining solution containing 40% v/v methanol and 10% v/v acetic acid solution was added to the gel and the gel mixture was slowly shaken for 30 to 60 min. The solution was replenished several times until background of the gel is fully destained. The gel was transferred to ultra-pure water, and the stained gel was stored in chilled temperatures (4oC) in a closed cleaned container.

Image Analysis of the Gel

The polyacrylamide gel was taken for scanned using Molecular Imager (Gel Doc XR+). The gel was placed properly on the scanner to have balance image. The bubbles under the gel must be removed to avoid the interfering background during scanning the gel. The image obtained was further analyses in Image Lab 6.0 (Bio-Rad). The colour of the image must be inverted prior to select the bands to enhance the intensity of bands for selection. The bands were selected manually to ensure all bands were quantified without missing the small intensity bands. The molecular weight of protein bands was added manually by referring to the manual of standard used during electrophoresis.

Statistical Analysis

Data collected in this study was calculated from three replications and analyses using SPSS (Statistical Package for the Social Sciences) version 19.0. One-way ANOVA test was used to compare differences in the mean values of the total soluble protein concentration and protein yield from different samples. This was followed by a Tukey post-hoc analysis to determine their differences. A significant difference was considered at the level of $p < 0.05$.

RESULTS AND DISCUSSIONS

Total Soluble Protein Concentration and Yield

Total soluble protein concentration was highly depended on the method of extraction. The pre-selection of extraction method from a few researchers have been analyses and phenol with ammonium acetate precipitation according to Bocian et al. (2019) has been selected as the best method. Total soluble protein concentration of the different samples as quantified by Bradford protein assay protocols are shown in Table 1. Overall, the honey from A. dorsata, A. melifera and H. itama content significantly higher protein concentration (0.92 ug/uL, 0.86 ug/uL and 0.82 ug/uL) compares to the honey from G. thoracica (0.73 ug/uL). Protein concentration of A. melifera honey shows that there was no significant difference (p> 0.05) between A. dorsata and H. itama honey samples, as well as A. dorsata and H. itama no significant difference (p> 0.05) was found. Only G. thoracica has shown significantly different (p<0.05) of protein concentration compared to the others honey samples.

The obvious difference of protein yield may be due to the environment of the honeybee that produce honey as the protein is belongs to the plant or bee origin (Lee et al., 1998). Bocian et al. (2019) reported that acacia honey (A. melifera) have protein concentration ranging from 4 to 10 µg/µL which is much higher from the results of this study. Previous study, Ajitha Nath et al. (2018) reported that stingless bee honey showed higher protein content compared with Apis honey. This dissimilarity may happen due to the differences in method of extraction, geological and other factors. Ultimately, the extraction method in this study was applicable in further MS/MS-based peptide identifications for protein discovery.

Table 1. Total soluble protein concentrations and yields

Sample Honey	Protein Concentrations (ug/uL)	Protein Yields (ug/g)
A. melifera	0.86 ± 0.06[a]	150.50 ± 20.94[a]
A. dorsata	0.92 ± 0.03[a]	169.75 ± 16.82[a]
G. thoracica	0.73 ± 0.05[b]	127.75 ± 19.69[b]
H. itama	0.82 ± 0.07[a]	143.58 ± 21.27[a]

Note: values presented as means (n=3) ± standard deviation and different letters in the column indicate significant differences (p<0.05).

SDS-PAGE Profile

A series of electrophoretic separations were performed for every honey variety and isolation method. SDS-PAGE protein bands of A. mellifera, A. dorsata, G. thoracica and H. itama was shown in Figure 2. The protein fractions are identified in terms of their molecular weights referring to the protein marker. The results shown various bands and intensity from each of honey samples. Overall, A. mellifera only shown 5 bands, was distributed ranging in size from 250 kDa to 50 kDa. Meanwhile, A. dorsata shown 7 bands, was distributed ranging in size from 250 kDa to 20 kDa. The number of protein bands of Apini tribe was less compared to Meliponini tribe. Number of protein band for G. thoracica and H. itama was 10 and 11 bands, respectively. All the bands were ranging in size from 250 kDa to 10 kDa. Obtained protein bands clearly show that the honey proteome varies depending on the type of honey (Bocian et al., 2019). Marshall & Williams (1987) reported of 19 bands in

SDS-PAGE of potential proteins extracted from honey. All the 19 bands were ranging in size from 500 kDa to below 10 kDa. Honey contains proteins in minute quantities and several enzymes are important components of honey such as a-glucosidase, b-glucosidase, amylase and glucose oxidase (Won et al 2008). (Baroni et al 2002) reported that pollen from different plants could be distinguished by SDS-PAGE and pollen proteins can be used as chemical markers for honey floral classification.

Figure 2. The protein molecular weight profile of honey protein based on SDS-PAGE analysis (12.5% acrylamide gel).
Note: From left the first lane is protein marker, followed by protein from A. mellifera, A. dorsata, G. thoracica and H. itama.

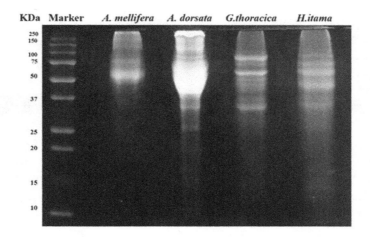

The SDS-PAGE pattern of each type of honey were analyses separately to view the details of the bands obtained using Image Lab 6.0 (Bio-Rad). The intensity of protein band and molecular weight can be visualized for each of the proteins. Figures 3, 4, 5 and 6 shows the densitometric scans protein bands from A. mellifera, A. dorsata, G. thoracica and H. itama. This image analysis will further characterize to identify the differences of the bands for each honey sample. Figure 3 shows the intensity of the protein bands from A. mellifera accumulated between 250 kDa to 50 kDa by referring to the protein marker. Protein band at 56.3 kDa was shown the highest intense compared to the other bands as seen in the densitometric scan images of protein profile for A. mellifera. Overall, 5 total bands that are shown in the Figure 3 indicated that at least 5 types of protein will be existing in A. mellifera honey. Di Girolamo et al. (2012) reported that 7 proteins have been identified from acacia honey and all the seven proteins were of animal origin (A. mellifera). The goal of that study was to find proteins of plant origin that would help to categorize the various honeys according to the type of flora from which they originated. Unfortunately, the finding of their result doesn't show any identified protein from plant origin.

Figure 3. Electropherogram and densitometric scan images of protein profile for A. mellifera.

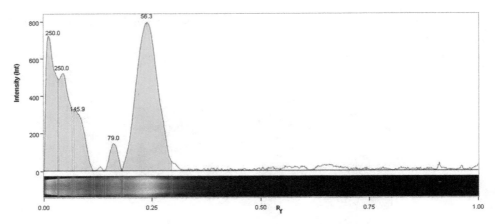
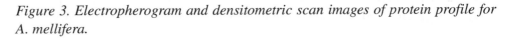

Electropherogram and densitometric scan images of protein profile from A. dorsata (Tualang honey) was shown in Figure 4. The Major intensity of the band was at 50.3 kDa and displayed 7 bands all together. In addition to variations in the geographical and floral origin of honey, differences in molecular weight of these major proteins in honey were noted as Apis cerana possessed a molecular weight of 56 kDa and A. mellifera had a molecular weight of 59 kDa (Lee et al., 1998). From these results, we determined differences in the molecular weight of two major proteins derived from different bee species. The variability may due to the Tualang honey (A. dorsata) is a Malaysian multifloral jungle honey meanwhile Manuka honey (A. mellifera) is a New Zealand and/or Australian monofloral honey.

Figure 5 shows the electropherogram and densitometric scan images of protein profile from G. thoracica. Results were displayed 9 distinctive bands from protein honey of G. thoracica with the highest intensity at molecular weight 85.6 kDa and 59.4 kDa. The low intensity of band at molecular weight 14.6 kDa. Studied by (Ramón-Sierra et al 2015) reported that protein honey from Trigona spp. (stingless bee) contained 14 bands with different intensity and the molecular weight of the proteins was between 9.2 to 128.0 kDa. Only three bands share similar molecular weight when comparing Trigona spp. with A. mellifera in their study. In this study, the results show that G. thoracica honey has various molecular weight bands compared to A. mellifera and A. dorsata honey which mean that there was more different type of proteins present in G. thoracica honey.

Figure 4. Electropherogram and densitometric scan images of protein profile for A. dorsata.

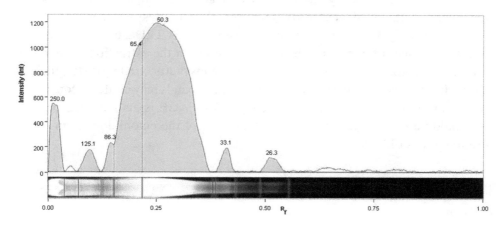

Figure 5. Electropherogram and densitometric scan images of protein profile for G. thoracica.

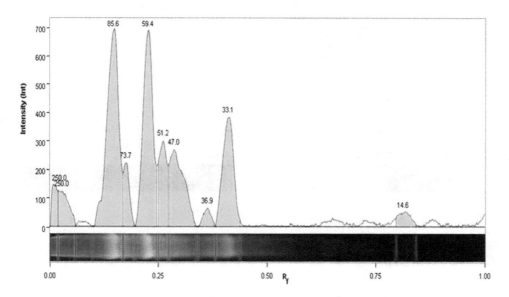

H. itama displayed the higher intensity of band at molecular weight 46.2 kDa and lowest intensity at 12.3 kDa. Figure 6 shown electropherogram and densitometric scan images of protein profile from H. itama. It contained 11 distinctive bands and resulted in higher amounts of bands compared among other honey protein samples. This result also display less number of bands compare with the profile of SDS-PAGE reported by Ramón-Sierra et al. (2015) that reveals 14 bands. Meanwhile, Ajitha Nath et al. (2018) reported that stingless bee samples of their studies show only 3 bands with molecular weight of 53.2 kDa, 65.5 kDa and 93.5 kDa, less number of bands may be due to different of the extraction method use. Thus, honeys protein extracts have distinctive bands that accurately distinguish the entomological origin of the sample. The pattern of bands may could be considered as protein markers for each type of honey. SDS–PAGE is an analytical method that could be used to determine the authenticity of the entomological origin of different types of honeys.

Figure 6. Electropherogram and densitometric scan images of protein profile for H.itama.

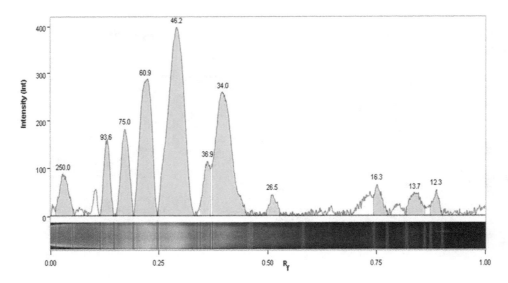

CONCLUSION

This study concluded that the concentration of protein in honey varies among each other. The results show that Meliponini tribe has more types of protein exist in their honey. Accordingly, to the results which mentioned for the first time using

SDS-PAGE for some honeys that comes from both tribes located in Malaysia, we can conclude that, researchers can use the protein profile in many aspects such as in authentication study of the unique protein that specifically available in particular types of honey where this uniqueness can be used as Unique Kelulut Factor (UKF). We manage to extract the protein from both tribe and further protein identification will be done by in-solution digestion and tandem mass spectrometry analysis (MS/MS-based peptide identifications). Thus, this study as an initial step in discovering the proteome profile of honeys in Malaysia.

CONFLICT OF INTEREST

We declare that there is no conflict of interest.

ACKNOWLEDGMENT

This work was funded by the Fundamental Research Grant Scheme (FRGS/1/2019/WAB11/UMT/02/2) and Naluri Pantas Sdn. Bhd. The authors would like to extend their gratitude to everyone who helped in the article writing process.

REFERENCES

Ajitha Nath, K. G. R., Jayakumaran Nair, A., & VS, S. (2018). *Comparison of proteins in two honey samples from Apis and stingless bee.* Academic Press.

Alimentarius, C. (2001). Revised Codex Standard for Honey. Codex Stan. 12-1981, Rev. 1 (1987). World Health Organization.

Ávila, S., Hornung, P. S., Teixeira, G. L., Malunga, L. N., Apea-Bah, F. B., Beux, M. R., Beta, T., & Ribani, R. H. (2019). Bioactive compounds and biological properties of Brazilian stingless bee honey have a strong relationship with the pollen floral origin. *Food Research International, 123*, 1–10. doi:10.1016/j.foodres.2019.01.068 PMID:31284956

Azevedo, M. S., Seraglio, S. K. T., Rocha, G., Balderas, C. B., Piovezan, M., Gonzaga, L. V., de Barcellos Falkenberg, D., Fett, R., de Oliveira, M. A. L., & Costa, A. C. O. (2017). Free amino acid determination by GC-MS combined with a chemometric approach for geographical classification of bracatinga honeydew honey (Mimosa scabrella Bentham). *Food Control, 78*, 383–392. doi:10.1016/j.foodcont.2017.03.008

Baroni, M. V., Chiabrando, G. A., Costa, C., & Wunderlin, D. A. (2002). Assessment of the floral origin of honey by SDS-page immunoblot techniques. *Journal of Agricultural and Food Chemistry*, *50*(6), 1362–1367. doi:10.1021/jf011214i PMID:11879003

Bocian, A., Buczkowicz, J., Jaromin, M., Hus, K. K., & Legáth, J. (2019). An effective method of isolating honey proteins. *Molecules (Basel, Switzerland)*, *24*(13), 2399. doi:10.3390/molecules24132399 PMID:31261846

Bradford, M. M. (1976). A rapid and sensitive method for the quantitation of microgram quantities of protein utilizing the principle of protein-dye binding. *Analytical Biochemistry*, *72*(1–2), 248–254. doi:10.1016/0003-2697(76)90527-3 PMID:942051

Carpentier, S. C., Witters, E., Laukens, K., Deckers, P., Swennen, R., & Panis, B. (2005). Preparation of protein extracts from recalcitrant plant tissues: An evaluation of different methods for two-dimensional gel electrophoresis analysis. *Proteomics*, *5*(10), 2497–2507. doi:10.1002/pmic.200401222 PMID:15912556

Corradini, E., Schmidt, P. J., Meynard, D., Garuti, C., Montosi, G., Chen, S., Vukicevic, S., Pietrangelo, A., Lin, H. Y., & Babitt, J. L. (2010). BMP6 treatment compensates for the molecular defect and ameliorates hemochromatosis in Hfe knockout mice. *Gastroenterology*, *139*(5), 1721–1729. doi:10.1053/j.gastro.2010.07.044 PMID:20682319

Di Girolamo, F., D'Amato, A., & Righetti, P. G. (2012). Assessment of the floral origin of honey via proteomic tools. *Journal of Proteomics*, *75*(12), 3688–3693. doi:10.1016/j.jprot.2012.04.029 PMID:22571915

Ibáñez, A., Riveros, R., Hurtado, E., Gleichgerrcht, E., Urquina, H., Herrera, E., Amoruso, L., Reyes, M. M., & Manes, F. (2012). The face and its emotion: Right N170 deficits in structural processing and early emotional discrimination in schizophrenic patients and relatives. *Psychiatry Research*, *195*(1–2), 18–26. doi:10.1016/j.psychres.2011.07.027 PMID:21824666

Jamnik, P., Raspor, P., & Javornik, B. (2012). A proteomic approach for investigation of bee products: Royal jelly, propolis and honey. *Food Technology and Biotechnology*, *50*(3), 270–274.

Kruger, N. J. (2002). Detection of polypeptides on immunoblots using enzyme-conjugated or radiolabeled secondary ligands. In *The Protein Protocols Handbook* (pp. 405–414). Springer. doi:10.1385/1-59259-169-8:405

Laemmli, U. K. (1970). Cleavage of structural proteins during the assembly of the head of bacteriophage T4. *Nature*, *227*(5259), 680–685. doi:10.1038/227680a0 PMID:5432063

Lee, D.-C., Lee, S.-Y., Cha, S.-H., Choi, Y.-S., & Rhee, H.-I. (1998). Discrimination of native bee-honey and foreign bee-honey by SDS-PAGE. *Korean Journal of Food Science Technology*, *30*(1), 1–5. doi:10.9721/KJFST.2017.49.1.1

Marshall, T., & Williams, K. M. (1987). Electrophoresis of honey: Characterization of trace proteins from a complex biological matrix by silver staining. *Analytical Biochemistry*, *167*(2), 301–303. doi:10.1016/0003-2697(87)90168-0 PMID:2450485

Naila, A., Flint, S. H., Sulaiman, A. Z., Ajit, A., & Weeds, Z. (2018). Classical and novel approaches to the analysis of honey and detection of adulterants. *Food Control*, *90*, 152–165. doi:10.1016/j.foodcont.2018.02.027

Pedreschi, R., Hertog, M., Lilley, K. S., & Nicolai, B. (2010). Proteomics for the food industry: Opportunities and challenges. *Critical Reviews in Food Science and Nutrition*, *50*(7), 680–692. doi:10.1080/10408390903044214 PMID:20694929

Queiroz-Junior, C. M., Silveira, K. D., de Oliveira, C. R., Moura, A. P., Madeira, M. F. M., Soriani, F. M., Ferreira, A. J., Fukada, S. Y., Teixeira, M. M., Souza, D. G., & da Silva, T. A. (2015). Protective effects of the angiotensin type 1 receptor antagonist losartan in infection-induced and arthritis-associated alveolar bone loss. *Journal of Periodontal Research*, *50*(6), 814–823. doi:10.1111/jre.12269 PMID:25753377

Ramón-Sierra, J. M., Ruiz-Ruiz, J. C., & de la Luz Ortiz-Vázquez, E. (2015). Electrophoresis characterization of protein as a method to establish the entomological origin of stingless bee honeys. *Food Chemistry*, *183*, 43–48. doi:10.1016/j.foodchem.2015.03.015 PMID:25863608

Rao, P. V., Krishnan, K. T., Salleh, N., & Gan, S. H. (2016). Biological and therapeutic effects of honey produced by honey bees and stingless bees: A comparative review. *Revista Brasileira de Farmacognosia*, *26*(5), 657–664. doi:10.1016/j.bjp.2016.01.012

Seraglio, S. K. T., Silva, B., Bergamo, G., Brugnerotto, P., Gonzaga, L. V., Fett, R., & Costa, A. C. O. (2019). An overview of physicochemical characteristics and health-promoting properties of honeydew honey. *Food Research International*, *119*, 44–66. doi:10.1016/j.foodres.2019.01.028 PMID:30884675

Šimúth, J., Bíliková, K., Kováčová, E., Kuzmová, Z., & Schroder, W. (2004). Immunochemical approach to detection of adulteration in honey: Physiologically active royal jelly protein stimulating TNF-α release is a regular component of honey. *Journal of Agricultural and Food Chemistry*, *52*(8), 2154–2158. doi:10.1021/jf034777y PMID:15080614

Won, S.-R., Lee, D.-C., Ko, S. H., Kim, J.-W., & Rhee, H.-I. (2008). Honey major protein characterization and its application to adulteration detection. *Food Research International*, *41*(10), 952–956. doi:10.1016/j.foodres.2008.07.014

Zulkhairi Amin, F. A., Sabri, S., Mohammad, S. M., Ismail, M., Chan, K. W., Ismail, N., Norhaizan, M. E., & Zawawi, N. (2018). Therapeutic properties of stingless bee honey in comparison with european bee honey. *Advances in Pharmacological Sciences*, *2018*, 2018. doi:10.1155/2018/6179596 PMID:30687402

Chapter 7

Dehydration Treatment Effect on the Physicochemical Properties and Microbial Population of Stingless Bee Honey From Three Different Species

Mannur Ismail Shaik
Universiti Malaysia Terengganu, Malaysia

Noor Zulaika Zulkifli
Universiti Malaysia Terengganu, Malaysia

Jaheera Anwar Sayyed
Universiti Malaysia Terengganu, Malaysia

John Sushma Nannepaga
Sri Padmavati Mahila Visvavidyalayam, India

Guruswami Gurusubramanian
Mizoram University, India

Shamsul Bahri Abd Razak
Universiti Malaysia Terengganu, Malaysia

ABSTRACT

Honey is a natural product produced from the nectar of a variety of plants by stingless bees. Honey has been utilized for nutritional food for ages, and in recent years,

DOI: 10.4018/978-1-6684-6265-2.ch007

stingless bee honey has been exploited as a food supplement for excellent health, cosmetic maintenance, and culinary enjoyment. Stingless bee honey has a higher moisture content (30-40%) and acidity than honeybee honey due to the presence of organic acid, mineral, and other trace components. Honey's moisture content is a key aspect that affects its stability and shelf life. The current study aimed to accesses the quality of dehydrated stingless bee honey from three different species namely, Heterotrigona itama, Geniotrigona thoracica, and Tetrigona apicalis. The dehydration treatment of T1 (20% moisture content at 60°C in 6 hours) and T2 (15% moisture content at 60°C in 8 hours) honey samples were subjected to physicochemical properties and microbial population studies.

INTRODUCTION

Stingless bee is known as a non-stinger bee and there are about 500 species of stingless bee that could be found across the globe. They are distributed in Latin America (Melipona, Tetragonisca, Scaptotrigona and Plebeia), the mainland of Australia (Tetragonula), Africa (Meliponula) and tropical parts of Asia (Lepidotrigona, Tetrigona, Homotrigona, Lisotrigona) (Nordin et al., 2018). In tropical countries like Malaysia, Thailand, Mexico, Venezuela, Brazil, and Australia stingless beekeeping practice is a better-known tradition. In Malaysia, more than 30 species of stingless bees locally known as "Kelulut" were found (Shamsudin et al., 2019). *Geniotrigona thoracica, Heterotrigona itama, Lepidotrigona Terminata, Tetragonula fuscobalteata,* and *Tetraponera laeviceps* are the most popular species for bee raising and economic value (Kelly et al., 2014). However, stingless bee honey has good quality and is reported to have antitumoral, antimicrobial, and antioxidant activities (Lani et al., 2017).

Honey is a complex food substance with around 200 distinct ingredients, including fructose, glucose, water, proteins, vitamins, minerals, polyphenolic chemicals, and plant derivatives (Nolan et al., 2019). Stingless bee honey is an astounding 'miracle liquid' with countless medicinal properties for various diseases such as gastroenteritis, and cataracts, as well as for wound healing (Rosli et al., 2020). It has been long recognized that the quality of honey is influenced by seasonal variations, post-harvest handling of honey, and storage condition. Due to the enormous demand for stingless bee honey and its therapeutic potential, quality and authenticity are still key considerations in its consumption and marketing (Gela et al., 2021). Because of its short shelf life, stingless bee honey has a limited global distribution (Moo-Huchin et al., 2015).

The physical and chemical properties of honey are qualitative parameters that are essential to determine its suitability for processing into a commercial product. The physicochemical characteristics include pH, electrical conductivity (EC), ash content, colour intensity and 5-hydroxymethylfurfural (HMF) which plays a vital role in the self-life of honey. In addition, honey's physical and chemical qualities are further influenced by variances in flora, meteorological circumstances, and geographical area. Furthermore, the microbial community of bacteria and fungus was an important factor in honey quality (Wanjai et al., 2012). Fresh harvested stingless bee honey has high moisture content and acidity due to the presence of organic acids, minerals, and other compounds (Özbalci et al., 2013). High moisture levels can result in the fermentation of honey which can affect the nutrients, active ingredients, and quality. Furthermore, one of the procedures used to make stingless bee honey is dehydration, which is a part of honey processing that is necessary to lower the moisture content of honey. It is important to reduce the moisture level of honey into its stable condition in the range of £20% (Singh & Singh, 2018).

Heterotrigona itama (H. itama), Tetrigona apicalis (T. apicalis), and *Geniotrigona thoracica (G. thoracica)* are three stingless bee species that have diverse characteristics. H. itama is the smallest stingless and Apis bee species, with a body size of 3.0-7.5 mm and a mostly blackish body with one weak tooth on the mandible (Pangestika et al., 2017). *T. apicalis* has a black head with fine white to yellowish hairs covering the frons, and thickening yellowish-white hair towards the antennal sockets and clypeus (Jongjitvimol & Wattanachaiyingcharoen, 2006). The overall length measurement of 5.53-7.75 mm indicates a higher body size. It has enormous honey and bee bread pots (Abdullah et al., 2017). *G. throcica* is one of the largest stingless bees found in Indo-Malayan areas. The size of the body varies from 6.67 mm to 10.80 mm (Saufi & Thevan, 2015). The hives of *G. thoracica* are generally found in trunk hollows, tree brunches, and rock crevices. Finally, the findings will aid in the establishment of stingless bee honey quality standards and characterization, especially in the identification of Malaysian stingless bee honey. The purpose of this study was to assess the effectiveness of dehydration treatment in reducing the water content of honey produced by three species: *H. itama, T. apicalis,* and *G. thoracica* (see Figure 1).

Figure 1. Stingless bee species; a) Heterotrigona itama; b) Tetrigona apicalis; c) Geniotrigona thoracica

MATERIALS AND METHODS

Sampling Site

Stingless bee honey samples of three different species namely, *H. itama, G. thoracica* and, *T. apicalis* species were collected from the Taman Kuantan, Pahang. The present study was conducted in the Crop Science laboratory of the Faculty of Fisheries and Food Science, University Malaysia Terengganu (See Figure 2).

Figure 2. Location of stingless bee honey sampling

Dehydration of Honey

Stingless bees honey sample (200 ml) was filled into the tray before running the dehydration experiment using the oven. The 60 °C temperatures were employed to conduct dehydration treatment, which reduced the water content to 20% in 6 hours (T1) and 15% in 12 hours (T2). The sample was checked every half-hour to ensure that the water content was properly reduced. A handheld digital refractometer was used to measure sample moisture (Kruss, HRH30). After treatment, the honey was placed into the refractometer plate and covered to record the reading. This procedure was performed three times to ensure that the results were accurate.

PHYSICOCHEMICAL PROPERTIES

pH

The pH of the honey sample was determined using a pH meter (Model 3505, Jenway, UK) according to (Belay et al., 2013). 10 g of the honey sample was measured and mixed well with 75 mL of sterile distilled water until a homogenous solution was obtained. After calibrating the pH meter with a standard buffer solution with pH of 4, 7, and 10, the glass electrode was gently inserted into the honey solution. The readings were recorded, and the procedure was repeated three times to achieve an accurate result.

Electrical Conductivity

The electrical conductivity (EC) of the honey sample was determined using the conductivity meter (H1 98311). The solution was prepared by dissolving 2 g of honey in 10mL of distilled water. After calibrating the conductivity meter, the conductance cell was suspended in the honey solution and the readings were recorded (Chuttong et al., 2016). The procedure was repeated three times for each sample to get an accurate result.

Ash

The ash content of the honey sample was determined according to the method by Chuttong et al. (2016). The dehydrated honey samples were weighed and recorded. 5 g of honey sample was placed in a muffle furnace at 550 °C temperature overnight. The weight of the cooled samples was measured and recorded. The percentage (%) of ash content was determined by the following formula.

$$\% \, Ash \, Content = \frac{Weight \, of \, Ash}{Weight \, of \, Sample} \times 100 \tag{1}$$

Colour

The colour intensity was measured using the method of (Pontis et al., 2014). 5g of each of the honey samples was dissolved in 10ml deionized water and gently mixed with a vortex mixer. About 2ml of the solution was transferred into curvets and the absorbance was measured using a spectrophotometer at 635nm. The recorded absorbance was matched with the Pfund scale (Table 1. and Figure 3).

Table 1. Colour Designation of Honey

USDA Colour Standard Designation	Colour Range Pfund Scale (mm)	Sample Results Range
Water White	£ 8	0 – 0.094
Extra White	> 8 and £ 17	0.094 – 0.189
White	> 17 and £ 34	0.189 – 0.378
Extra Light Amber	> 34 and £ 50	0.378 – 0.595
Light Amber	> 50 and £ 85	0.595 – 1.389
Amber	> 85 and £ 114	1.389 – 3.008
Dark Amber	> 114	> 3.008

Source: United States Department of Agriculture (2020)

Figure 3. Chart for Identification of Honey Colour
Source: http://uaglobalinc.com/acacia-honey/

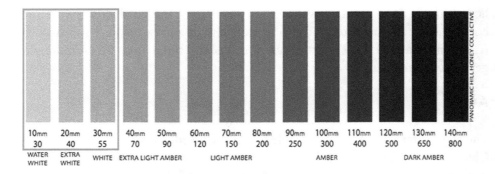

Hydroxymenthylfurfural (HMF)

The concentration of Hydroxymenthylfurfural (HMF) was measured by using reverse-phase RP-HPLC equipped with UV detection (Ciulu et al., 2013). 1g of honey was weighed accurately and dissolved in 5ml of water until achieving a homogenous solution. The solution was mixed well by using a vortex and the syringe and pre-mounted 0.45µm membrane filter were used for transferring the sample solutions into the vials and the sample solution was ready for chromatography.

MICROBIAL POPULATION ANALYSIS

The microbial population of stingless bee honey was determined by Nutrient Agar (NA) media for bacteria and Potato Dextrose Agar (PDA) media for fungi. A single colony was obtained by using the streak plate technique followed by the pour plate and the spread plate method. A colony-forming unit (CFU) was used to estimate the number of viable bacterial and fungal cells in the sample (Sanders, 2012).

Plate Count

One milliliter of stingless bee honey sample was measured and poured into a sterilized McCartney bottle. Then, 9 ml of sterile distilled water was added and mixed thoroughly. From the homogeneous suspension, the serial dilution was carried out up to 10-4 dilutions. Aliquots (0.1ml) of 10-2, 10-3, and 10-4 were poured into the Nutrient Agar (NA) and Potato Dextrose Agar (PDA) media and cultured triplicate by the spread plate method. The PDA plates were incubated for 7 days at 28 °C, whereas NA plates were incubated for 24 hours at 37 °C (Thawai et al., 2004). The number of CFU per milliliter of a sample was calculated by using the following formula.

$$CFU/ml\ in\ Original\ Sample = \frac{Colonies\ Count \times Dilution\ Factor}{Volume\ of\ Culture\ Plate} \qquad (2)$$

Statistical Analysis

All measurable data on the differentiation physicochemical properties and microbial population between three species of stingless bee honey were subjected to statistical analysis using Genstat 19th edition. The analysis used was ANOVA; specifically, Two-Way-ANOVA. The levels of significance with $P<0.05$ were considered.

RESULTS AND DISCUSSION

pH

Naturally, the stingless bee honey has sourness in taste and a distinct aroma due to its acidic nature ranging from 3.2 to 4.5. The current study reveals that the pH values of three different species of stingless bee honey are within the range of 3.12 to 3.403 for honey without dehydration treatment. The mean pH values from *G. thoracica* (3.12) exhibit more acidic followed by *T. apicalis* (3.22), and *H. itama* (3.403). After dehydration treatment, the pH values increased in all species respective to decreased moisture content. The pH values for *H. itama, T. apicalis,* and *G. thoracica* were recoded as 3.47 and 3.49, 3.29 and 3.31, 3.14 and 3.15 after the T1 and T2 treatments respectively. According to the current findings, G. thoracica is more acidic than the other two species, *T. apicalis* and *H. itama*. In H. itama species, show that the pH value of the control is more acidic compared to the two treatments which are T1 and T2. This result is also the same with the other two species, *T. apicalis,* and *G. thoracica* (Figure 4).

Figure 4. pH values of three species of stingless bee honey with dehydration treatment

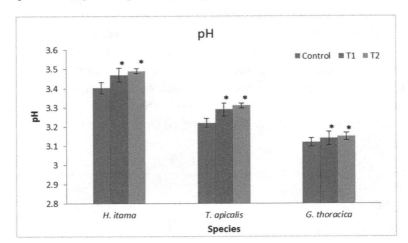

Honey's acidity is caused by environmental factors, mineral content, and the varieties of plant nectar gathered by stingless bee workers (Lee et al., 2008). However, a study conducted by de Sousa et al. (2016) revealed that the acidic value was corresponding to the balance of organic acid present in the honey, where the pH value varies and depends on the floral origin and the bee species. According to the result reported by da Silva et al. (2013), the lower pH of honey inhibits the growth of microorganisms. A possible explanation for this finding is that the longer heating time may volatize the organic acid present in stingless bee honey, as well as a specific acid produced by the dehydration method, resulting in a drop in acidity and, as a result, a rise in pH. (Araújo, 2001). Furthermore, the Malaysian Standard has established a pH range of 2.5 to 3.8 for stingless bee honey as acceptable.

Electrical Conductivity

The electrical conductivity (EC) of stingless bee honey samples was determined using the conductivity meter and the result is depicted in Figure 5.

Figure 5. Electrical Conductivity of three species of stingless bee honey with dehydration treatment

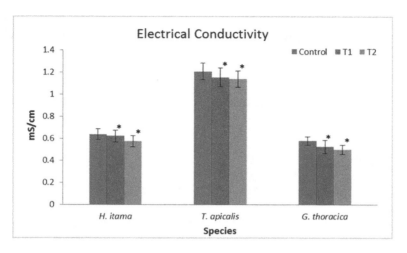

The results have shown that the highest electrical conductivity for *T. apicalis* of 1.21 ms/cm before treatment and significantly reduced values of 1.15 and 1.14 ms/cm after the T1 and T2 treatments respectively. The lowest electrical conductivity values were observed in *G. thoracica*, and a significant difference was found without dehydration treatment (0.58 ms/cm) and with the dehydration treatments (0.52 and 0.50 ms/cm) T1 and T2 respectively. A similar trend was found in *H. itama* shows moderate EC values compared to the other two species, without treatment 0.64 ms/cm and with treatment T1 0.62 ms/cm and T2 0.57 ms/cm. The current data evidenced by previous electrical conductivity studies reported by Chuttong et al. (2016) and Suntiparapop et al. (2012) ranged from 0.32 to 3.10 ms/cm in the stingless bee honey.

Furthermore, the electrical conductivity (EC) of the stingless bee honey is directly related to its mineral concentration, salts, organic acids, and proteins as well as an indication of the origin and source of nectar (Solayman et al., 2016). As a result, this parameter has been utilized as a honey quality indicator to aid in the identification and differentiation of floral honeys (Karabagias et al., 2014). This finding demonstrates that the duration of dehydration therapy has minimal influence on EC. The EC value should not exceed 0.8 ms/cm, according to the International Honey Commission (IHC). However, Malaysia Standard, suggests the unsuitability of this parameter for a stingless bee honey standard (Nordin et al., 2018).

Ash

The ash content of stingless bee honey was depicted in Figure 6. *T. apicalis* with the T2 treatment had the greatest ash concentration, at 1.31 percent, followed by T1 and without treatment, at 1.29 percent and 1.14 percent, respectively.

Figure 6. Ash content of three species of stingless bee honey with dehydration treatment

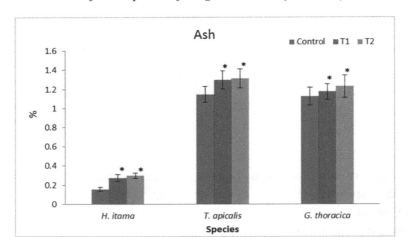

The lowest ash content was shown in *H. itama* honey at 0.15% before treatment and 0.27% and 0.30% after the dehydration treatment T1 and T2 respectively. The *G. thoracica* honey showed moderated levels of ash content, which is 1.12% without treatment, 1.17% for T1, and 1.23% for T2 treatment. According to Santos et al. (2014), the ash content of stingless bee honey following dehydration treatment ranged from 0.02 percent to 0.19 percent, depending on the kind of flower. Moreover, the ash content was recorded in the range of 0.13% to 0.66% for *T. thorasica* and *H. itama* stingless bee honey (Fatima et al., 2018). However, the ash content, on the other hand, illustrates the abundance of mineral content in honey sources, which is primarily impacted by nectar botanical origin, location, bee species, processing and handling, and represents the overall mineral content in honey (Gela et al., 2021). The ash level of honey samples was measured in this study, and significant variations were found across species and dehydration treatment.

Colour

The colour intensity of all honey samples was categorized based on the absorbance measurement and the values are depicted in Figure 7. The honey colour intensity was changed from light amber to dark amber with dehydration treatment, as classified by the Pfund scale. The current results demonstrated that in colour intensity of honey significant differences between without and with dehydration treatment were observed. Without dehydration treatment, all three stingless bee species had the lowest colour intensity of honey (0.42, 0.75, and 0.67). However, colour intensity of honey was gradually increased with the dehydration treatment T1 (0.53, 1.02, 0.9) and T2 (0.56, 1.04, 094) for *G. thoracica, T. apicalis* and *H. itama* respectively. Honey's colour is determined by minerals, pollen, phenolic concentration, storage period, and temperature (Solayman et al., 2016). 5-hydroxymethylfurfural impacted the colour intensity of honey samples (Can et al., 2015).

Moreover, the colour intensity of *T. thorasica* and *H. itama* stingless bee honey was found to be within the range of 0.23 to 1.05 (Fatima et al., 2018). The darkening of honey is highly dependent on the dehydration temperature and at 70 °C the colour intensity of Kelulut (*H. itama*) honey noticeably increased compared to 40 and 55 °C (Yap et al., 2019). Cui et al. (2008) found that changes in honey colour are minimal at temperatures between 30 and 50°C, while Turkmen et al. (2006) found that colour changes dramatically at temperatures between 50 and 70 °C. The Maillard reaction or fructose caramelization might explain why the colour intensity

Figure 7. Colour intensity of three species of stingless bee honey with dehydration treatment

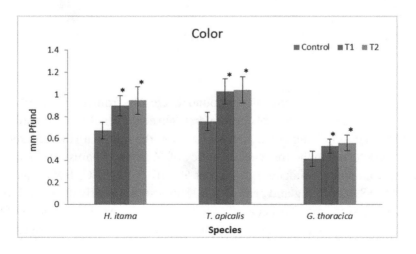

of the stingless bee honey sample rose as the dehydration duration increased (de Almeida-Muradian et al., 2013).

Hydroxymenthylfurfural (HMF)

The HMF content of stingless bee honey was measured using RP-HPLC and values are represented in Figure 8. The lowest HMF content of honey samples without dehydration treatment was noticed as 7.8 mg/kg, 2.3 mg/kg, and 15.7 mg/kg in *H. itama, T. apicalis*, and *G. thoracica* respectively. The elevated levels of HMF content were found with dehydration treatment of honey samples. The highest HMF content levels were found as 2510, 1012, and 477 mg/kg with T2 treatment followed by 683, 258, and 241 mg/kg with T1 treatment in *T. apicalis, G. thoracica* and *H. itama* respectively.

Figure 8. Hydroxymenthylfurfural (HMF) values of three species of stingless bee honey with dehydration treatment

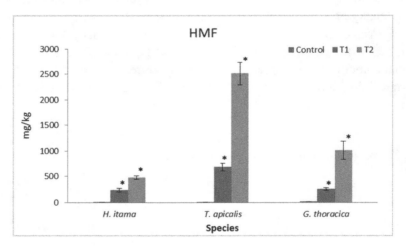

HMF is a furan group chemical compound that results from the Maillard reaction, which involves the dehydration of carbohydrates in acidic conditions, notably during thermal treatments (Zhang et al., 2012). However, the minimal HMF content (0.11 mg/kg) was detected in the fresh gelam honey of *H. itama* (Shamsudin et al., 2019). According to a study conducted by Yap et al. (2019) at 70°C, HMF content was increased to 189.15 mg/kg and peaked at 60 h and subsequently decreased to 24.69 mg/kg at 84 h. Moreover, Tosi et al. (2008) showed that the HMF content of honey increased more significantly at higher heating temperatures and longer heating duration. The heat treatment of honey causes increased levels of HMF content (Samborska & Czelejewska, 2014). HMF is widely known as a parameter of freshness in honey and tends to increase during the processing or aging of the product. For commercial export, the HMF should be less than 40 mg/kg (Codex Alimentarius).

Bacteria

The microbial population of the honey sample was determined using the plate count method and the bacterial count was shown in Figure 9. The highest bacterial count was found in without dehydration treatment samples of three different species as *H. itama* (2.7x107) *T. apicalis* (2.0x107), and *G. thoracica* (2.1x107). The decreased bacterial count was observed in the T1 treatment of honey from all three species *H. itama* (7.8x106), *T. apicalis* (5.4x106) and *G. thoracica* (6.7x106). The lowest bacterial count was found in the T2 treatment of honey from all three species *H. itama* (4.4x106), *T. apicalis* (3.8x106), and *G. thoracica* (4.2x106). The prolonged dehydration period was linked to a lower bacterial count in the current investigation. In stingless bee honey, dehydration is linked to the destruction of microorganisms. Dehydration treatment is associated with destroying microbes in stingless bee honey as indicated by the significant difference between without treatment and treatment groups.

Figure 9. Bacteria count of three species of stingless bee honey with dehydration treatment

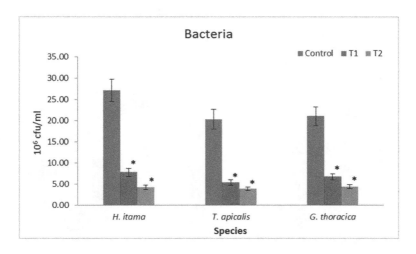

Yap et al. (2019) suggested that the dehydration process can reduce the water activity of honey to a level safe in terms of microbial growth. Honey contains antibacterial characteristics that can slow or stop the growth of a variety of microorganisms (Adenekan et al., 2010). Microbial contamination during or after processing can cause deterioration or the persistence of certain bacteria in honey, in addition to contamination at the source of the honey's indigenous microflora (Akharaiyi & Lawal, 2016). Molds, which are regularly discovered in honey, may survive but not reproduce, therefore high activity is generally linked to recent environmental or equipment contamination during processing (Finola et al., 2007).

Fungi

The plate count method was used to determine the fungal count of honey samples, with the findings shown in Figure 10. Similar to the bacterial count, the fungal count was also highest in the without dehydration treatment honey samples and lowest in the T2 treatment samples. The highest fungal count was found in the non-dehydrated control samples of three different species as *H. itama* (2.2x107) *T. apicalis* (1.6x107), and *G. thoracica* (1.7x107). The decreased fungal count was observed in the T1 treatment of honey from all three species *H. itama* (7.4x106), *T. apicalis* (5.3x106), and *G. thoracica* (7.1x106). The lowest bacterial count was found in the T2 treatment of honey from all three species *H. itama* (3.5x106), *T. apicalis* (2.9x106), and *G. thoracica* (3.3x106).

Figure 10. Fungi count of three species of stingless bee honey with dehydration treatment

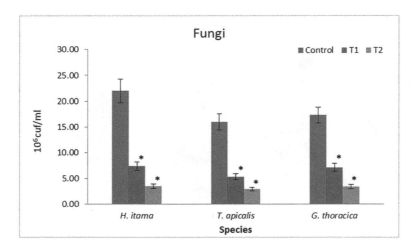

Pollen, the digestive tracts of honey bees, dust, nectar, and other main sources (during harvest) can all include bacteria that are difficult to manage (Pucciarelli et al., 2014). The drop in fungal populations is linked to a decrease in overall microbial activity. Besides the presence of yeast in stingless bee honey might also contribute to the increase of the total microbial activity in stingless bee honey (Wanjai et al., 2012). Finola et al. (2007) found that a high microbe population can cause honey to become more acidic. As a result, the data showed that a decrease in free acidity was linked to a decrease in yeast population.

CONCLUSION

The present findings revealed and compared the different dehydration treatments of honey harvested by three species of stingless bees such as *G. thoracica, T. apicalis* and *H. itama*. Two dehydration treatments help in preventing undesirable honey fermentation and increase honey's shelf life by lowering the honey's moisture content. Physicochemical parameters including pH, ash content, colour intensity, and HMF content were increased, but electrical conductivity was decreased. Except for the high levels of HMF content, all other physicochemical parameters are within acceptable levels. When stingless bee honey was dehydrated, the bacterial and fungal population of the honey was reduced. In conclusion, the dehydrated treatments improve the quality of stingless bee honey by reducing moisture content and minimizing microbial population. However, the HMF content rapidly increased with dehydration treatment which was beyond the accepted range. Therefore, with a lower HMF level, effective dehydration treatment procedures must be established.

ACKNOWLEDGMENT

We would like to thank the reviewers for all their constructive comments and suggestions.

REFERENCES

Abdullah, M., Bakhtan, M. A. H., & Mokhtar, S. A. (2017). Number Plate Recognition Of Malaysia Vehicles Using Smearing Algorithm. *Science International (Lahore), 29*(4), 823–827.

Adenekan, M. O., Amusa, N. A., Lawal, A. O., & Okpeze, V. E. (2010). Physico-chemical and microbiological properties of honey samples obtained from Ibadan. *Journal of Microbiology and Antimicrobials*, *2*(8), 100–104.

Akharaiyi, F. C., & Lawal, H. A. (2016). Physicochemical analysis and mineral contents of honey from farmers in western states of Nigeria. *J. Nat. Sci. Res*, *6*, 78–84.

Araújo, J. M. A. (2001). Food chemistry: Theory and practice. Federal University of Viçosa Press.

Belay, A., Solomon, W. K., Bultossa, G., Adgaba, N., & Melaku, S. (2013). Physicochemical properties of the Harenna forest honey, Bale, Ethiopia. *Food Chemistry*, *141*(4), 3386–3392. doi:10.1016/j.foodchem.2013.06.035 PMID:23993497

Can, Z., Yildiz, O., Sahin, H., Turumtay, E. A., Silici, S., & Kolayli, S. (2015). An investigation of Turkish honeys: Their physico-chemical properties, antioxidant capacities and phenolic profiles. *Food Chemistry*, *180*, 133–141. doi:10.1016/j.foodchem.2015.02.024 PMID:25766810

Chuttong, B., Chanbang, Y., Sringarm, K., & Burgett, M. (2016). Physicochemical profiles of stingless bee (Apidae: Meliponini) honey from South east Asia (Thailand). *Food Chemistry*, *192*, 149–155. doi:10.1016/j.foodchem.2015.06.089 PMID:26304332

Ciulu, M., Farre, R., Floris, I., Nurchi, V. M., Panzanelli, A., Pilo, M. I., Spano, N., & Sanna, G. (2013). Determination of 5-hydroxymethyl-2-furaldehyde in royal jelly by a rapid reversed phase HPLC method. *Analytical Methods*, *5*(19), 5010–5013. doi:10.1039/c3ay40634b

Cui, Z.-W., Sun, L.-J., Chen, W., & Sun, D.-W. (2008). Preparation of dry honey by microwave–vacuum drying. *Journal of Food Engineering*, *84*(4), 582–590. doi:10.1016/j.jfoodeng.2007.06.027

da Silva, I. A. A., da Silva, T. M. S., Camara, C. A., Queiroz, N., Magnani, M., de Novais, J. S., Soledade, L. E. B., de Oliveira Lima, E., de Souza, A. L., & de Souza, A. G. (2013). Phenolic profile, antioxidant activity and palynological analysis of stingless bee honey from Amazonas, Northern Brazil. *Food Chemistry*, *141*(4), 3552–3558. doi:10.1016/j.foodchem.2013.06.072 PMID:23993520

de Almeida-Muradian, L. B., Stramm, K. M., Horita, A., Barth, O. M., da Silva de Freitas, A., & Estevinho, L. M. (2013). Comparative study of the physicochemical and palynological characteristics of honey from M elipona subnitida and A pis mellifera. *International Journal of Food Science & Technology*, *48*(8), 1698–1706. doi:10.1111/ijfs.12140 de Sousa, J. M. B., de Souza, E. L., Marques, G., de Toledo Benassi, M., Gullón, B., Pintado, M. M., & Magnani, M. (2016). Sugar profile, physicochemical and sensory aspects of monofloral honeys produced by different stingless bee species in Brazilian semi-arid region. *Lebensmittel-Wissenschaft + Technologie, 65,* 645–651. doi:10.1016/j.lwt.2015.08.058

Fatima, I. J., AB, M. H., Salwani, I., & Lavaniya, M. (2018). Physicochemical characteristics of malaysian stingless bee honey from trigona species. *IIUM Medical Journal Malaysia, 17*(1), 187–191. doi:10.31436/imjm.v17i1.1030

Finola, M. S., Lasagno, M. C., & Marioli, J. M. (2007). Microbiological and chemical characterization of honeys from central Argentina. *Food Chemistry, 100*(4), 1649–1653. doi:10.1016/j.foodchem.2005.12.046

Gela, A., Hora, Z. A., Kebebe, D., & Gebresilassie, A. (2021). Physico-chemical characteristics of honey produced by stingless bees (Meliponula beccarii) from West Showa zone of Oromia Region, Ethiopia. *Heliyon, 7*(1), 1–7. doi:10.1016/j. heliyon.2020.e05875 PMID:33506124

Jongjitvimol, T., & Wattanachaiyingcharoen, W. (2006). Pollen food sources of the stingless bees Trigona apicalis Smith, 1857, Trigona collina Smith, 1857 and Trigona fimbriata Smith, 1857 (Apidae, Meliponinae) in Thailand. *Tropical Natural History, 6*(2), 75–82.

Karabagias, I. K., Badeka, A., Kontakos, S., Karabournioti, S., & Kontominas, M. G. (2014). Characterisation and classification of Greek pine honeys according to their geographical origin based on volatiles, physicochemical parameters and chemometrics. *Food Chemistry, 146,* 548–557. doi:10.1016/j.foodchem.2013.09.105 PMID:24176380

Kelly, N., Farisya, M. S. N., Kumara, T. K., & Marcela, P. (2014). Species Diversity and External Nest Characteristics of Stingless Bees in Meliponiculture. *Pertanika. Journal of Tropical Agricultural Science, 37*(3), 293–298.

Lani, M. N., Zainudin, A. H., Razak, S. B. A., Mansor, A., & Hassan, Z. (2017). Microbiological quality and pH changes of honey produced by stingless bees, Heterotrigona itama and Geniotrigona thoracica stored at ambient temperature. *Malaysian Applied Biology, 46*(3), 89–96.

Lee, H., Churey, J. J., & Worobo, R. W. (2008). Antimicrobial activity of bacterial isolates from different floral sources of honey. *International Journal of Food Microbiology, 126*(1–2), 240–244. doi:10.1016/j.ijfoodmicro.2008.04.030 PMID:18538876

Moo-Huchin, V. M., González-Aguilar, G. A., Lira-Maas, J. D., Pérez-Pacheco, E., Estrada-León, R., Moo-Huchin, M. I., & Sauri-Duch, E. (2015). Physicochemical properties of Melipona beecheii honey of the Yucatan Peninsula. *Journal of Food Research*, *4*(5), 25–32. doi:10.5539/jfr.v4n5p25

Nolan, V. C., Harrison, J., & Cox, J. A. G. (2019). Dissecting the antimicrobial composition of honey. *Antibiotics (Basel, Switzerland)*, *8*(4), 1–26. doi:10.3390/antibiotics8040251 PMID:31817375

Nordin, A., Sainik, N. Q. A. V., Chowdhury, S. R., Saim, A., & Idrus, R. B. H. (2018). Physicochemical properties of stingless bee honey from around the globe: A comprehensive review. *Journal of Food Composition and Analysis*, *73*, 91–102. doi:10.1016/j.jfca.2018.06.002

Özbalci, B., Boyaci, İ. H., Topcu, A., Kadılar, C., & Tamer, U. (2013). Rapid analysis of sugars in honey by processing Raman spectrum using chemometric methods and artificial neural networks. *Food Chemistry*, *136*(3–4), 1444–1452. doi:10.1016/j.foodchem.2012.09.064 PMID:23194547

Pangestika, N. W., Atmowidi, T., & Kahono, S. (2017). Pollen load and flower constancy of three species of stingless bees (Hymenoptera, Apidae, Meliponinae). *Tropical Life Sciences Research*, *28*(2), 179–187. doi:10.21315/tlsr2017.28.2.13 PMID:28890769

Pontis, J. A., da Costa, L. A. M. A., da Silva, S. J. R., & Flach, A. (2014). Color, phenolic and flavonoid content, and antioxidant activity of honey from Roraima, Brazil. *Food Science and Technology (Campinas)*, *34*(1), 69–73. doi:10.1590/S0101-20612014005000015

Pucciarelli, A. B., Schapovaloff, M. E., Kummritz, S., Señuk, I. A., Brumovsky, L. A., & Dallagnol, A. M. (2014). Microbiological and physicochemical analysis of yateí (Tetragonisca angustula) honey for assessing quality standards and commercialization. *Revista Argentina de Microbiologia*, *46*(4), 325–332. doi:10.1016/S0325-7541(14)70091-4 PMID:25576417

Rosli, F. N., Hazemi, M. H. F., Akbar, M. A., Basir, S., Kassim, H., & Bunawan, H. (2020). Stingless bee honey: Evaluating its antibacterial activity and bacterial diversity. *Insects*, *11*(8), 1–13. doi:10.3390/insects11080500 PMID:32759701

Samborska, K., & Czelejewska, M. (2014). The influence of thermal treatment and spray drying on the physicochemical properties of Polish honeys. *Journal of Food Processing and Preservation*, *38*(1), 413–419. doi:10.1111/j.1745-4549.2012.00789.x

Sanders, E. R. (2012). Aseptic Laboratory Techniques: Plating Methods (2022). *Journal of Visualized Experiments*, *63*(63), e3064. doi:10.3791/3064 PMID:22617405

Saufi, N. F. M., & Thevan, K. (2015). Characterization of nest structure and foraging activity of stingless bee, Geniotrigona thoracica (Hymenopetra: Apidae; Meliponini). *Jurnal Teknologi*, *77*(33), 69–74.

Shamsudin, S., Selamat, J., Sanny, M., Abd. Razak, S.-B., Jambari, N. N., Mian, Z., & Khatib, A. (2019). Influence of origins and bee species on physicochemical, antioxidant properties and botanical discrimination of stingless bee honey. *International Journal of Food Properties*, *22*(1), 239–264. doi:10.1080/10942912 .2019.1576730

Singh, I., & Singh, S. (2018). Honey moisture reduction and its quality. *Journal of Food Science and Technology*, *55*(10), 3861–3871. doi:10.100713197-018-3341-5 PMID:30228384

Solayman, M., Islam, M. A., Paul, S., Ali, Y., Khalil, M. I., Alam, N., & Gan, S. H. (2016). Physicochemical properties, minerals, trace elements, and heavy metals in honey of different origins: A comprehensive review. *Comprehensive Reviews in Food Science and Food Safety*, *15*(1), 219–233. doi:10.1111/1541-4337.12182 PMID:33371579

Suntiparapop, K., Prapaipong, P., & Chantawannakul, P. (2012). Chemical and biological properties of honey from Thai stingless bee (Tetragonula leaviceps). *Journal of Apicultural Research*, *51*(1), 45–52. doi:10.3896/IBRA.1.51.1.06

Thawai, C., Tanasupawat, S., Itoh, T., Suwanborirux, K., & Kudo, T. (2004). Micromonospora aurantionigra sp. nov., isolated from a peat swamp forest in Thailand. *Actinomycetologica*, *18*(1), 8–14. doi:10.3209aj.18_8

Tosi, E., Martinet, R., Ortega, M., Lucero, H., & Ré, E. (2008). Honey diastase activity modified by heating. *Food Chemistry*, *106*(3), 883–887. doi:10.1016/j. foodchem.2007.04.025

Turkmen, N., Sari, F., Poyrazoglu, E. S., & Velioglu, Y. S. (2006). Effects of prolonged heating on antioxidant activity and colour of honey. *Food Chemistry*, *95*(4), 653–657. doi:10.1016/j.foodchem.2005.02.004

Wanjai, C., Sringarm, K., Santasup, C., Pak-Uthai, S., & Chantawannakul, P. (2012). Physicochemical and microbiological properties of longan, bitter bush, sunflower and litchi honeys produced by Apis mellifera in Northern Thailand. *Journal of Apicultural Research*, *51*(1), 36–44. doi:10.3896/IBRA.1.51.1.05

Yap, S. K., Chin, N. L., Yusof, Y. A., & Chong, K. Y. (2019). Quality characteristics of dehydrated raw Kelulut honey. *International Journal of Food Properties*, *22*(1), 556–571. doi:10.1080/10942912.2019.1590398

Zhang, Y., Song, Y., Zhou, T., Liao, X., Hu, X., & Li, Q. (2012). Kinetics of 5-hydroxymethylfurfural formation in chinese acacia honey during heat treatment. *Food Science and Biotechnology*, *21*(6), 1627–1632. doi:10.100710068-012-0216-9 PMID:31807335

Chapter 8
Microbiological Diversity and Properties of Stingless Bee Honey

Amir Izzwan Zamri
Universiti Malaysia Terengganu, Malaysia

Nor Hazwani Mohd Hasali
Universiti Malaysia Terengganu, Malaysia

Muhammad Hariz Mohd Hasali
Universiti Malaysia Terengganu, Malaysia

Tuan Zainazor Tuan Chilek
Universiti Malaysia Terengganu, Malaysia

Fisal Ahmad
Universiti Malaysia Terengganu, Malaysia

Mohd Khairi Mohamed Zainol
Universiti Malaysia Terengganu, Malaysia

ABSTRACT

The study was to compare and evaluate the performance of stingless bee honey (Heterotrigona itama spp.) with ordinary honey in terms of proximate composition as a comparison. Both honeys have shown diverse application and importance either traditionally and scientifically. However, due to the heightened interest on stingless bee honey, antimicrobial tests were also performed to determine the inhibition activity of stingless bee honey against food-borne pathogens using agar well diffusion assay. All three honey samples showed very good inhibitory activities (measured by

DOI: 10.4018/978-1-6684-6265-2.ch008

inhibition zone) against Salmonella typhimurium (25-33 mm), Escherichia coli (17-33 mm), Pseudomonas aeruginosa (15-25 mm), and Staphylococcus aureus (25-29 mm). As for resistance to bile salts, pH tolerance was done and indicated the Lactic acid bacteria was able to survive the human digestive system. The haemolytic study shows that the LAB used was not virulent when introduced to red blood cells, which is important for any bacterium to be classified as safe.

INTRODUCTION

Just by mentioning honey, our thoughts will be directed to the difficult process of obtaining the honey, whether it is a Tualang honey or a stingless bee honey. The challenge of obtaining the honey is what makes it one of the most valuable commodities. Around the world, the use of bee honey is very widespread, especially in traditional medicine and is used as a method to replace the use of sugar that is widely associated with diabetes. As a result of the increase in consumption and high demand, artificial bee honey is made using sugar and a mixture of other ingredients which can lead to health problems. However, as a result of the advances in beekeeping programs and the improvement of the industry, this problem is slowly diminishing.

In general, the beekeeping system in Malaysia, whether the common honeybees or stingless bee, has been progressing and is increasingly in line with more productive neighboring countries such as Thailand and Vietnam which utilizes intensive bee keeping systems. Therefore, various parties have conducted several attempts to promote the beekeeping industry to ensure increased honey production to the Malaysian market in recent years. This is carried out through attempts by a more structurally and modern system and by identifying the variety of bees and suitable areas for the bee keeping project. Demand for honey slowly began to increase since the beginning of 1984. This in turn have a positive effect and raised the honey industry to a higher platform. The honey industry was further strengthened and intensified due to the overwhelming demand as more benefits of the honey was discovered and published.

According to Ndubisi et al. (2008), several government agencies have taken steps to ensure honey production in Malaysia increases such as the Department of Agriculture (DOA), Malaysian Rubber Research Institute (RRIM), Rubber Industry Small Development Authority (RISDA) Institute Malaysian Agricultural Research and Development (MARDI), Universiti Putra Malaysia (UPM) and Universiti Malaysia Terengganu (UMT). All these agencies have run special area extension services for bee breeding to ensure more efficient production of honey resources.

This effort makes Malaysia a country rich in natural honey resources that have benefits especially for global human health. Throughout research and development combined with traditional knowledge it was discovered that the most common use of honey is to treat stomach discomfort, cough, tonsillitis, sore throat, ulcers, and as a wound dressing (Abd Jalil et al., 2017).

Honey is a natural food that is very rich in nutritional value. Honey consists of a complex chemical composition that provides many benefits to humans (Firdaus et al., 2018). In general, honey is a functional food that has been known for generations and is used to treat wounds, ulcers, sunburn, and eye, throat and intestinal infections. In 1892, the first study of the antimicrobial properties of honey was conducted which found that it has broad -spectrum antimicrobial properties against gram -positive and negative bacteria including Escherichia coli, Pseudomonas aeruginosa, Klebsiella pneuomniae, Staphylococcus aureus, Bacillus subtilis and Listeria monocytogens (Laallam et al., 2015). This show the deep interest in the nutritional composition and benefits of honey which dates back throughout the centuries.

TUALANG HONEY

Tualang honey originates from bees that build nests in Tualang trees or scientifically known as Koompassia excelsa. Koompassia excelsa is considered one of the most prominent trees in the tropical rainforests and among the tallest trees in the world. Tualang tree can be found in the rainforests of Sumatra, Borneo, Southern Thailand, and Peninsular Malaysia. Tualang trees are known by different names depending on areas such as Mengaris in Brunei and Sabah, Tualang in Peninsular Malaysia, Sialang in Indonesia, and Tapang in Sarawak. Tualang trees can grow to a height of more than 80 meters and one of the tallest tualang trees in Malaysia in Tawau Hills Park, Sabah reaches a height of up to 85 meters. This makes it stand out among other trees. Bees make the shade of this Tualang tree as a nesting place due to the high morphological condition of the tree, smooth and free from predation.

Tualang honey that has been taken from the Tualang tree is considered by locals in Malaysia as one of the most valuable honey because the honey itself is obtained from various types of flowers in tropical rainforests below the tree canopy and is rich in nutrients. Tualang honeycombs can be up to 25 feet above the tree and each hive can house more than 30000 bees hence making it a difficult process in obtaining the honey. Each Tualang tree can house up to more than 100 nests and can produce more than 450 kg of honey (Ahmed & Othman, 2013). In addition, honey is often classified and named according to the geographical location where it is produced, the way it is obtained and processed (Chin & Sowndhararajan, 2020). In Malaysia, particularly in the West Coast of Malaysia, especially in Perak, Negeri Sembilan, Johor, Melaka and Selangor.most of the bee honey is obtained from the bee from the species Apis Dorsata, Apis Cerana, and Apis Mellifera.

STINGLESS BEE

Kelulut bee (Trigona spp.) or also known as stingless bee (stingless bee) includes about 130 species worldwide. Trigona spp is the largest genus found in Neotropics (Mexico to Argentina), the Indo-Australia region to India (Sri-Lanka to Taiwan), Australia and Southern Indonesia (Choudhari et al., 2012). However, in Malaysia, only about 17 to 32 species of stingless bees are popular for honey production and pollination activities in fruit plantation areas (Norowi et al., 2010; Salim et al., 2012). Recently, many stingless bees beekeeping areas have been established, especially in Kelantan, Terengganu and Pahang.

There are five popular species known as Trigona thoracica (Genio Trigona), Trigona itama, (HeteroTrigona), Trigona terminate (Lepido Trigona), Trigona scintillans (LisoTrigona) and Trigona laeviceps (Tetragonula) which are abundant especially in bee farms in Kelantan (Kelly et al., 2014)., T. itama in particular is highly favored by bee farmers and this species accounts for 83.2% of the total colonies on the farm, followed by T. thoracica. Kelulut bees are very small social insects (in size) in comparison with other type of honey producing species that live in permanent colonies (Michener, 2000) usually in tree cavities, logs or even cracks. The diameter of the cavity varies with the type of nesting site. Most colonies of Trigona spp. construct the inlet tube from the resin and the newly constructed one is soft and then becomes darker and hardens due to maturation. Food storage zones are classified into honey zones and pollen zones. When compared to regular bee honey, stingless bee honey is more liquid, darker in color and has a sharper taste due to its higher acid content (da Silva et al., 2016; Halawani & Shohayeb, 2011). In some countries new methods have been utilized to facilitate stingless bee keeping such as the use of coconut shells that are attached to the wall by using mud as a tool for keeping bees which in turn will facilitate honey extraction. In some other parts of the world, the colonies are breed in boxes attached to log which mimic the actual condition in the wild.

Kelulut honey is highly valued in Ethiopia for example which is used to treat common diseases because it is believed to be stronger than ordinary honey. According to Andualem (2014), the antimicrobial activity of Trigona honey against the pathogens tested was greater than that of other types of honey. In addition, Trigona laeviceps bees found in Thailand, produce honey with antimicrobial activity against several types of bacteria namely E. coli and S. aureus and also the fungus Aspergillus Niger, as well as two types of yeast (Auriobasidium pullulans and Candida albicans) (Chanchao, 2009).

BASIC USES OF HONEY IN GENERAL

Honey is commonly known to have many nutrients and is effective in treating several diseases such as coughs, wounds, itchy skin caused by osmotic pressure from sugar content, pH, and the presence of good bacteria. Most types of honey produce hydrogen peroxide when diluted due to the activation of the enzyme glucose oxidase, which oxidizes glucose to gluconic acid and hydrogen peroxide. Hydrogen peroxide is a major ingredient that contributes to the antimicrobial activity found in honey and different concentrations of this compound in various honeys produce different antimicrobial effects (Nolan et al., 2019). However, it is determined by the habitat and the roaming habits of the bee (Taormina et al., 2001).

MORPHOLOGICAL COMPARISON BETWEEN MADU KELULUT AND TUALANG

The composition and physicochemical properties of honey are constantly changing depending on the weather, season and flower source around the beehive (El Sohaimy et al., 2015). In some instances, raw honey sometimes does not comply to the classification for example the Malaysia Food Law due to the variation caused particularly in the moisture content (maximum of 20%) which is dictated by the season (particularly the rainy season). This in turn require some intervention in technology in order to make the honey produced comply and be standardised. According to Özbalci et al. (2013), kelulut bee honey is a valuable bee product from Trigona spp. and is quite different in color, taste and viscosity compared to the honey produced by Tualang bees. The difference between Kelulut honey from Tualang honey is the high moisture content and acidity due to the presence of acids, minerals and other compounds. Apart from that, the honey storage containers between Tualang and Kelulut are also not the same with each other. Ordinary bees store honey in honeycombs while stingless bees store their honey in honey pots which both contribute to the difference in composition (see Figure 1). Table 1 shows the nutritional composition of kelulut honey.

Table 1. Physicochemical parameters for raw stingless bee honey

Parameters	Stingless Bee Honey	Ordinary Honey
pH	3.01-3.05	
Moisture (%)	21.3 – 31.7	19-29
Minerals (%)	0.04-0.19	0.16-1.2
Protein (%)	0.34-0.69	0.4-0.9
Fat (%)	0	0
Carbohydrate (%)	67.0 – 77.6	70-80
Energy (kcal/100g)	276- 317	280-320

MICROBIAL DIVERSITY IN STINGLESS BEE HONEY

Bee (wheather ordinary bee or stingless bee) has its own microbial content which contributes to the quality and morphology of the honey. Several types of bacteria have been isolated from honey and show antimicrobial activity against both gram-negative and positive bacteria as well as putrefaction bacteria (Aween et al., 2010; Ibarguren et al., 2010; Lee et al., 2008). Most of the antimicrobial activity in honey is due to the presence of organic acids and bacteriocin produced by several types of complex lactic acid bacteria (LAB), which have activities that can inhibit the growth of gram -positive and negative bacteria (Mohamed Mustafa Aween et al., 2012). Bacterocin is a protein -shaped toxin that inhibits or inhibits the growth of other bacteria. Several types of lactic acid bacteria especially from Lactobacillus spp have been successfully isolated from honey and studies show effects towards harmful Gram-positive and Gram positive bacteria or pathogens (Aween et al., 2010; Ibarguren et al., 2010; Klaenhammer, 1993; Lee et al., 2008). A study conducted by Hasali et al., (2015) showed that there are 5 types of lactic acid bacteria that have been successfully isolated from stingless bee honey from Kelantan and Terengganu which was defines as Lactobacillus brevis (lbr-42), Lactobacillus brevis (37901), Lactobacillus brevis (ATTC 367), Lactobacillus brevis (NJ42) and Lactobacillus brevis (KLDS).

Apart from that studies also show that lactic acid bacteria obtained from stingless bee honey can react and be inhibited by several types of antibiotics such as ampilicillin, erythromycin, chloramphenicol, tetracycline and kanamycin. The 5 types of lactic acid bacteria that were successfully isolated in the work by Hasali et al. (2015) also showed antagonistic effects against Bacillus cereus (ATTC11778), Salmonella Typhimurium, Pseudomonas aeruginosa, Escherichi coli (ATTC11775) and Staphylococcus aureus (ATTC259230). This shows the danger of bacteria found in honey can have a beneficial effect on bacterial infections and other problems caused by pathogenic bacteria. The ability of lactic acid bacteria can also be used in the field of food preservation where it can prevent damage to dairy food by unwanted bacteria. Due to the benefits seen in stingless bee honey there is a growing interest to study the content as well as lactic acid bacteria (LAB) from honey as it is recognized as safe (GRAS) and plays an important role in the fermentation and preservation of food (Yang et al., 2014). LAB is important for preserving the nutritional quality of raw materials through preservation, enhancing food flavoring and inhibiting bacteria and pathogenic bacteria on food products (Suhartatik et al., 2014).

Lactic acid bacteria (LAB) successfully isolated from natural honey from Malaysia, Libya, Saudi Arabia and Yemen isolated by heating were found to exhibit antifungal activity (Bulgasem et al., 2017). Lactic acid bacteria (LAB) are listed as one of the substances listed in the list ingredients Commonly Recognized as Safe (GRAS) by the United States Food and Drug Association (USFDA) based on their important role in food fermentation and probiotic properties in medical applications (Hoque et al., 2010). Lactic acid (LAB) bacteria can also produce soluble compounds such as lactic acid, diacetyl, acetaldehyde, and hydrogen peroxide that can also inhibit the growth of harmful microorganisms and pathogens (Allameh et al., 2012). This study is a continuation of previous studies published by Hasali et al. (2015).

Lactic acid bacteria (LABs) belong to the group of Gram-positive bacteria that produce lactic acid as their main fermentation product. Most of the antimicrobial activity of honey is due to metabolites or by -products produced by LAB, such as organic acids and bacteriocin (Aween et al., 2012). However, studies on the antimicrobial activity of honey from Meliponine bees, especially in Malaysia are still limited. The study of Hasali et al. (2018) also detailed some important effects of lactic acid bacteria isolated from kelulut honey. Among them are resistance to pH, resistance to bile salts and haemolytic activity.

pH is an important aspect in showing the resistance of a microbe, especially in the human digestive tract. Bacteria that are resistant to low pH have the potential to invade the digestive system and remain part of the microflora in the gut that will affect the process of digestion and digestion of food. It was found that microbes isolated from stingless bee honey can survive at pH 2 at 3 hours of incubation period. Which of these represents the time and conditions that will be passed during the process of digestion of food? The results of this study coincide with the study by Dunne et al. (2001) which stated that Lactobacillus bacteria need to have the ability to survive in acidic conditions (as low as pH 1 or 2) to enable it to exert a probiotic effect on the host.

As for resistance to bile salts, it is also part of the digestive system in addition to acids that need to be overcomed by lactic acid bacteria to enable them to live and thrive in the human digestive system. Shehata et al. (2016) stated that lactic acid bacteria should show resistance to bile salt concentrations at 0.3% (weight/volume).

The results of the study from Hasali et al. (2015) found that Lactobacillus isolated from stingless bee honey had resistance to bile salts at the required level which indicate that the bacteria will be able to survive the digestive system and flourish in the digestive system. Hemolytic activity shows virulent activity in a bacterial sepsis that will cause damage to red blood cells (hemolysis). Studies have found that lactic acid bacteria isolated from stingless bee honey do not show β-haemolytic properties. Only one strain was found to show α-haemolytic (partial hemolysis) characteristics. This is important as this will show that the Lactic acid bacteria will not able to cause damage in the digestive system as they flourish and become part of the digestive system microflora.

CONCLUSION

In general, kelulut honey is found to have a very good characteristics and is suitable for the human digestive system. It is important to ensure that lactic acid bacteria which is also a probiotic bacterium can pass through the human digestive system and live thriving in the human intestine well. Stingless bee honey is a very valuable gift. The nutritional and medicinal value of stingless bee honey has been widely proven through studies is very important to lift stingless bee honey to be an important food supplement. It can also be one of the methods that can be used and developed in alternative medicine that is more natural in nature that suits human nature.

ACKNOWLEDGMENT

Any acknowledgment to fellow researchers or funding grants should be placed within this section.

REFERENCES

Abd Jalil, M. A., Kasmuri, A. R., & Hadi, H. (2017). Stingless bee honey, the natural wound healer: A review. *Skin Pharmacology and Physiology*, *30*(2), 66–75. doi:10.1159/000458416 PMID:28291965

Ahmed, S., & Othman, N. H. (2013). Review of the medicinal effects of tualang honey and a comparison with manuka honey. *The Malaysian Journal of Medical Sciences: MJMS*, *20*(3), 6–13. PMID:23966819

Allameh, S. K., Daud, H., Yusoff, F. M., Saad, C. R., & Ideris, A. (2012). Isolation, identification and characterization of Leuconostoc mesenteroides as a new probiotic from intestine of snakehead fish (Channa striatus). *African Journal of Biotechnology*, *11*(16), 3810–3816.

Andualem, B. (2014). Physico-chemical, microbiological and antibacterial properties of Apis mellipodae and Trigona spp. honey against bacterial pathogens. *World Journal of Agricultural Sciences*, *10*(3), 112–120.

Aween, M. M. (2012). Evaluation on antibacterial activity of Lactobacillus acidophilus strains isolated from honey. *American Journal of Applied Sciences*, *9*(6), 807–817. doi:10.3844/ajassp.2012.807.817

Aween, M. M., Hassan, Z., Muhialdin, B. J., Eljamel, Y. A., Al-Mabrok, A. S. W., & Lani, M. N. (2012). Antibacterial activity of lactobacillus acidophilus strains isolated from honey marketed in malaysia against selected multiple antibiotic resistant (mar) gram-positive bacteria. *Journal of Food Science*, *77*(7), 364–371. doi:10.1111/j.1750-3841.2012.02776.x PMID:22757710

Aween, M. M., Zaiton, H., & Belal, J. (2010). Antimicrobial Activity of Lactic Acid Bacteria Isolated from Honey. *Proceedings of the International Symposium on Lactic Acid Bacteria (ISLAB'10), University Putra Malaysia*, 25–27.

Bulgasem, B. Y., Hassan, Z., Huda-Faujan, N., Omar, R. H. A., Lani, M. N., & Alshelmani, M. I. (2017). Effect of pH, heat treatment and enzymes on the antifungal activity of lactic acid bacteria against Candida species. *Malaysian Journal of Microbiology*, *13*(3), 195–202. doi:10.21161/mjm.89416

Chanchao, C. (2009). Antimicrobial activity by Trigona laeviceps (stingless bee) honey from Thailand. *Pakistan Journal of Medical Sciences*, *25*(3), 364–369.

Chin, N. L., & Sowndhararajan, K. (2020). A review on analytical methods for honey classification, identification and authentication. In *Honey Analysis-New Advances and Challenges*. IntechOpen. doi:10.5772/intechopen.90232

Choudhari, M. K., Punekar, S. A., Ranade, R. V., & Paknikar, K. M. (2012). Antimicrobial activity of stingless bee (Trigona sp.) propolis used in the folk medicine of Western Maharashtra, India. *Journal of Ethnopharmacology*, *141*(1), 363–367. doi:10.1016/j.jep.2012.02.047 PMID:22425711

da Silva, P. M., Gauche, C., Gonzaga, L. V., Costa, A. C. O., & Fett, R. (2016). Honey: Chemical composition, stability and authenticity. *Food Chemistry*, *196*, 309–323. doi:10.1016/j.foodchem.2015.09.051 PMID:26593496

Dunne, C., O'Mahony, L., Murphy, L., Thornton, G., Morrissey, D., O'Halloran, S., Feeney, M., Flynn, S., Fitzgerald, G., Daly, C., Kiely, B., O'Sullivan, G. C., Shanahan, F., & Collins, J. K. (2001). In vitro selection criteria for probiotic bacteria of human origin: Correlation with in vivo findings. *The American Journal of Clinical Nutrition*, *73*(2), 386s–392s. doi:10.1093/ajcn/73.2.386s PMID:11157346

El Sohaimy, S. A., Masry, S. H. D., & Shehata, M. G. (2015). Physicochemical characteristics of honey from different origins. *Annals of Agricultural Science*, *60*(2), 279–287. doi:10.1016/j.aoas.2015.10.015

Firdaus, A., Khalid, J., & Yong, Y. K. (2018). Malaysian Tualang honey and its potential anti-cancer properties: A review. *Sains Malaysiana*, *47*(11), 2705–2711. doi:10.17576/jsm-2018-4711-14

Halawani, E. M., & Shohayeb, M. M. (2011). Shaoka and Sidr honeys surpass in their antibacterial activity local and imported honeys available in Saudi markets against pathogenic and food spoilage bacteria. *Australian Journal of Basic and Applied Sciences*, *5*(4), 187–191.

Hasali, N. H. M., Zamri, A. I., Lani, M. N., Mubarak, A., & Suhaili, Z. (2015). Identification of lactic acid bacteria from Meliponine honey and their antimicrobial activity against pathogenic bacteria. *American-Eurasian Journal of Sustainable Agriculture*, *9*(6), 1–7.

Hasali, N. O. R. H., Zamri, A. I., Lani, M. N., Mubarak, A., Ahmad, F., & Chilek, T. Z. T. (2018). Physico-chemical analysis and antibacterial activity of raw honey of stingless bee farmed in coastal areas in Kelantan and Terengganu. *Malaysian Applied Biology*, *47*(4), 145–151.

Hoque, M. Z., Akter, F., Hossain, K. M., Rahman, M. S. M., Billah, M. M., & Islam, K. M. D. (2010). Isolation, identification and analysis of probiotic properties of Lactobacillus spp. from selective regional yoghurts. *World J Dairy Food Sci*, *5*(1), 39–46.

Ibarguren, C., Raya, R. R., Apella, M. C., & Audisio, M. C. (2010). Enterococcus faecium isolated from honey synthesized bacteriocin-like substances active against different Listeria monocytogenes strains. *Journal of Microbiology (Seoul, Korea)*, *48*(1), 44–52. doi:10.100712275-009-0177-8 PMID:20221729

Kelly, N., Farisya, M. S. N., Kumara, T. K., & Marcela, P. (2014). Species Diversity and External Nest Characteristics of Stingless Bees in Meliponiculture. *Pertanika. Journal of Tropical Agricultural Science*, *37*(3), 293–298.

Klaenhammer, T. R. (1993). Genetics of bacteriocins produced by lactic acid bacteria. *FEMS Microbiology Reviews*, *12*(1–3), 39–85. doi:10.1016/0168-6445(93)90057-G PMID:8398217

Laallam, H., Boughediri, L., Bissati, S., Menasria, T., Mouzaoui, M. S., Hadjadj, S., Hammoudi, R., & Chenchouni, H. (2015). Modeling the synergistic antibacterial effects of honey characteristics of different botanical origins from the Sahara Desert of Algeria. *Frontiers in Microbiology*, *6*, 1–12. doi:10.3389/fmicb.2015.01239 PMID:26594206

Lee, H., Churey, J. J., & Worobo, R. W. (2008). Antimicrobial activity of bacterial isolates from different floral sources of honey. *International Journal of Food Microbiology*, *126*(1–2), 240–244. doi:10.1016/j.ijfoodmicro.2008.04.030 PMID:18538876

Michener, C. D. (2000). *The bees of the world* (Vol. 1). Johns HopNins University Press.

Ndubisi, N. O., Malhotra, N. K., & Wah, C. K. (2008). Relationship marketing, customer satisfaction and loyalty: A theoretical and empirical analysis from an Asian perspective. *Journal of International Consumer Marketing*, *21*(1), 5–16. doi:10.1080/08961530802125134

Nolan, V. C., Harrison, J., & Cox, J. A. G. (2019). Dissecting the antimicrobial composition of honey. *Antibiotics (Basel, Switzerland)*, *8*(4), 1–26. doi:10.3390/antibiotics8040251 PMID:31817375

Norowi, M. H., Mohd, F., Sajap, A. S., Rosliza, J., & Suri, R. (2010). Conservation and sustainable utilization of stingless bees for pollination services in agricultural ecosystems in Malaysia. *Proceedings of International Seminar on Enhancement of Functional Biodiversity Relevant to Sustainable Food Production in ASPAC*, 1–11.

Özbalci, B., Boyaci, İ. H., Topcu, A., Kadılar, C., & Tamer, U. (2013). Rapid analysis of sugars in honey by processing Raman spectrum using chemometric methods and artificial neural networks. *Food Chemistry*, *136*(3–4), 1444–1452. doi:10.1016/j.foodchem.2012.09.064 PMID:23194547

Salim, H. M. W., Dzulkiply, A. D., Harrison, R. D., Fletcher, C., Kassim, A. R., & Potts, M. D. (2012). Stingless bee (Hymenoptera: Apidae: Meliponini) diversity in dipterocarp forest reserves in Peninsular Malaysia. *The Raffles Bulletin of Zoology*, *60*(1), 213–219.

Shehata, M. G., El Sohaimy, S. A., El-Sahn, M. A., & Youssef, M. M. (2016). Screening of isolated potential probiotic lactic acid bacteria for cholesterol lowering property and bile salt hydrolase activity. *Annals of Agricultural Science*, *61*(1), 65–75. doi:10.1016/j.aoas.2016.03.001

Suhartatik, N., Cahyanto, M. N., Rahardjo, S., Miyashita, M., & Rahayu, E. S. (2014). Isolation and identification of lactic acid bacteria producing [Beta] glucosidase from Indonesian fermented foods. *International Food Research Journal*, *21*(3), 973–978.

Taormina, P. J., Niemira, B. A., & Beuchat, L. R. (2001). Inhibitory activity of honey against foodborne pathogens as influenced by the presence of hydrogen peroxide and level of antioxidant power. *International Journal of Food Microbiology*, *69*(3), 217–225. doi:10.1016/S0168-1605(01)00505-0 PMID:11603859

Yang, S.-C., Lin, C.-H., Sung, C. T., & Fang, J.-Y. (2014). Antibacterial activities of bacteriocins: Application in foods and pharmaceuticals. *Frontiers in Microbiology*, *5*, 1–14.

Chapter 9
Morphometric Analysis in Stingless Bee (*Apidae meliponini*) Diversity

Suhaila Ab Hamid
Universiti Sains Malaysia, Malaysia

ABSTRACT

Insects occur in large numbers. Therefore, it is important to have a system to identify the different species of insects. Traditional morphological identification of insects requires an experienced entomologist while molecular techniques require laboratory expertise and involve substantial costs. Due to that, there has been a dramatic increase of studies using morphometric analysis in understanding the systematics, taxonomy, and diversity of stingless bees. Morphometric analysis is a powerful tool as it is effective with minimum technical experience. It is a simple technique because of the current availability of cheap computer technology equipped with software, and at the same time, this method preserves the physical integrity of the shape measured. Morphometric analysis makes it credible to recognise morphological disparity and lead ways to explore the causes, both within and between, the stingless bee populations.

INTRODUCTION

Insect pollinators are important in improving crop productivity and helping achieve optimum pollination during flowering. The evolution of self and maintenance of mixed mating systems is theoretically affected by pollination ecology (Brunet & Sweet, 2006). The stability, diversity, and function of natural and agricultural plant

DOI: 10.4018/978-1-6684-6265-2.ch009

communities are influenced by insect pollinators (Saunders, 2018). The quality and quantity of crop yields in the agricultural system are influenced by insect pollinators that provide the ecosystem service. In contrast, the natural system is influenced by plant reproduction and community (Saunders, 2018). The majority of the pollinating insects are from the order Hymenoptera.

In insect nomenclature, Hymenoptera is the third largest insect order and probably the most beneficial to humans of all insect orders. Like other hymenopterans, stingless bees are members of the class Insecta under the family Apidae and live in colonies (groups). The order of Hymenoptera is called "Hymen," which comes from the Greek word meaning god of marriage because the forewing and hind wings are joined together with small hooks (hamuli). There are two pairs of membranous wings (Plate 1) and three pairs of legs located at the thorax, divided into coxa, trochanter, femur, tibia, and metatarsus sections and claws (Plate 2). Most stingless bee workers have a modified hind legs structure and a corbicular (pollen basket) for collecting and transporting pollen and other materials. After a foraging activity, these pollen baskets are stuffed full of bright yellow or orange pollen.

Figure 1. The forewing of a stingless bee.

MORPHOMETRIC ANALYSIS IN STINGLESS BEE DIVERSITY

The world is facing climatic changes, which have altered certain insects' distribution and phenology (Batista et al., 2011). According to Lancaster et al. (2016), insects vary their morphological traits across climatic gradients. The variation in insect morphological traits may have benefits such as survival and enable it to colonize wide environmental tolerances (Bai et al., 2016). Different survival patterns have taken place in distinguishing insect species' resilience from different morph frequencies

(Yadav et al., 2018). Evidence suggests smaller stingless bee species have better behavioral proficiency maintenance than larger bees (Duell, 2018), which provides an important finding in stingless bees' persistence. Morphological similarities between species or cryptic species make morphological identification of stingless bees more difficult. Variations in the morphology of the insect typically intensify the identification process. Variation is a normal characteristic of any morphological element. Populations of the same species in different locations may encounter different ecological and climatic conditions, causing one or more characteristics to vary. Size variations also provide important clues about insect species taxonomy, anatomy and biology.

Figure 2. The leg (femur, tibia & metatarsus) of a stingless bee.

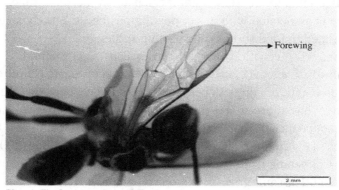

Plate 1: The forewing of stingless bee.

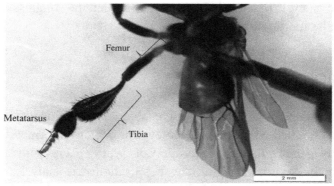

Plate 2: The leg (femur, tibia & metatarsus) of stingless bee.

The morphometric analysis is being used to solve a series of biological problems in studies of morphological variation of the genetic basis, insect growth, and insect evolution (Klingenberg 2002). According to Rohlf (1990), morphometrics is a method of analysis, quantitative description, variation, and interpretation of form in biology. This easy-to-adopt method could summarize the morphological data graphically and numerically and examine the relationships in many dimensions. The measurement of morphometric data can become more precise with the help of computer technology using various methods for analysis. Morphological characters that can be used in stingless bee morphometric study as species identification may include tibia, mandible, head, propodeum, mesoscutum, body, thorax, abdomen, wing venation, and wing length (Sakagami et al., 1990). Previously, the study showed morphometric analysis based on wings is useful for evaluating the population genetic divergence species of bees (Oleksa & Tofilski, 2015). It is essential for conservation as they seem less susceptible to environmental influence. At the same time, the stingless bee body size is associated with insect populations because body size differences are usually related to food sources and nesting strategies.

In addition, wing venation morphometrics can explain stingless bee species complexity and differences (Laksono et al., 2020). According to Hamid et al. (2016), the mean body length of stingless bees, Tetragonula laeviceps, in forest areas was slightly greater than those found in urban areas. The possible reason for this is probably the abundance of food sources from all flowering plants and fruit trees in the forest area. This finding, also supported by Kuberappa et al. (2005), reported that Trigona iridipennis from the hilly zone was bigger than in the urban area in Karnataka, India. The same goes for Trigona binghami in Thailand which was found bigger in a forest than in urban areas (Klakasikorn et al., 2005). It confirms that morphometric analysis has shown differences between the stingless bee, although they look similar morphologically.

CONCLUSION

Morphometric analysis can be used for characterizing and identifying species of stingless bees. This method can provide biology and ecology information on stingless bees, especially in unravelling the uncertainty of cryptic species with morphological changes caused by environmental influence. Morphometric analysis is a useful and reliable additional method for species discrimination on stingless bee diversity.

ACKNOWLEDGMENT

Any acknowledgment to fellow researchers or funding grants should be placed within this section.

REFERENCES

Bai, Y., Dong, J.-J., Guan, D.-L., Xie, J.-Y., & Xu, S.-Q. (2016). Geographic variation in wing size and shape of the grasshopper Trilophidia annulata (Orthoptera: Oedipodidae): morphological trait variations follow an ecogeographical rule. *Scientific Reports*, *6*(1), 1–15. doi:10.1038rep32680 PMID:27597437

Batista, E. C., Carvalho, L. R., Casarini, D. E., Carmona, A. K., Dos Santos, E. L., Da Silva, E. D., Dos Santos, R. A., Nakaie, C. R., Rojas, M. V. M., de Oliveira, S. M., Bader, M., D'Almeida, V., Martins, A. M., de Picoly Souza, K., & Pesquero, J. B. (2011). ACE activity is modulated by the enzyme α-galactosidase A. *Journal of Molecular Medicine*, *89*(1), 65–74. doi:10.100700109-010-0686-2 PMID:20941593

Brunet, J., & Sweet, H. R. (2006). Impact of insect pollinator group and floral display size on outcrossing rate. *Evolution; International Journal of Organic Evolution*, *60*(2), 234–246. doi:10.1111/j.0014-3820.2006.tb01102.x PMID:16610316

Duell, M. E. (2018). *Matters of size: behavioral, morphological, and physiological performance scaling among stingless bees (Meliponini)*. Arizona State University.

Hamid, S. A., Salleh, M. S., Thevan, K., & Hashim, N. A. (2016). Distribution and morphometrical variations of stingless bees (Apidae: Meliponini) in urban and forest areas of Penang Island, Malaysia. *J. Trop. Resour. Sustain. Sci*, *4*, 1–5.

Klakasikorn, A., Wongsiri, S., Deowanish, S., & Duangphakdee, O. (2005). New record of stingless bees (Meliponini: Trigona) in Thailand. *Tropical Natural History*, *5*(1), 1–7.

Klingenberg, C. P. (2002). Morphometrics and the role of the phenotype in studies of the evolution of developmental mechanisms. *Gene*, *287*(1–2), 3–10. doi:10.1016/S0378-1119(01)00867-8 PMID:11992717

Kuberappa, G. C., Mohite, S., & Kencharaddi, R. N. (2005). Biometrical variations among populations of stingless bee, Trigona iridipennis in Karnataka. *Indian Bee Journal*, *67*, 145–149.

Laksono, P., Raffiudin, R., & Juliandi, B. (2020). Stingless bee Tetragonula laeviceps and T. aff. biroi: Geometric morphometry analysis of wing venation variations. *IOP Conference Series. Earth and Environmental Science*, *457*(1), 12084. doi:10.1088/1755-1315/457/1/012084

Lancaster, L. T., Dudaniec, R. Y., Chauhan, P., Wellenreuther, M., Svensson, E. I., & Hansson, B. (2016). Gene expression under thermal stress varies across a geographical range expansion front. *Molecular Ecology*, *25*(5), 1141–1156. doi:10.1111/mec.13548 PMID:26821170

Oleksa, A., & Tofilski, A. (2015). Wing geometric morphometrics and microsatellite analysis provide similar discrimination of honey bee subspecies. *Apidologie*, *46*(1), 49–60. doi:10.100713592-014-0300-7

Rohlf, F. J. (1990). Morphometrics. *Annual Review of Ecology and Systematics*, 299–316.

Sakagami, S. F., Inoue, T., & Salmah, S. (1990). Stingless bees of central Sumatra. *Stingless Bees of Central Sumatra.*, 125–137.

Saunders, M. E. (2018). Insect pollinators collect pollen from wind-pollinated plants: Implications for pollination ecology and sustainable agriculture. *Insect Conservation and Diversity*, *11*(1), 13–31. doi:10.1111/icad.12243

Yadav, S., Stow, A. J., Harris, R. M. B., & Dudaniec, R. Y. (2018). Morphological variation tracks environmental gradients in an agricultural pest, Phaulacridium vittatum (Orthoptera: Acrididae). *Journal of Insect Science*, *18*(6), 1–13. doi:10.1093/jisesa/iey121 PMID:30508202

Chapter 10
Palynology of
Heterotrigona itama

Wan Noor Aida
Politeknik Jeli, Malaysia

Arifullah Mohammed
Universiti Malaysia Kelantan, Malaysia

Kumara Thevan
Universiti Malaysia Kelantan, Malaysia

ABSTRACT

Meliponiculture is the practice of handling the stingless bee for a lot of beneficial products including honey, bee bread, and propolis. Heterotrigona itama is one of the most cultured species by bee keepers in Malaysia. This research objective was pollen identification from pollen pot of H. itama. Five colonies of H. itama were observed from September 2014 until August 2015 for the foraging and pollen collection. Meanwhile, for palynology studies, the pollen sampling was done for four periods, which are September 2014, December 2014, March 2015, and June 2015. from three districts: Jeli, Kota Bharu, and Tanah Merah. One Way ANOVA was conducted, and results showed significant difference, p<0.05 for pollen area, pollen height of pot, pollen diameter, honey area, honey, honey diameter pot, honey height of pot, honey number of pot, and amount of honey. A total of 17,097 pollen were counted based on 66 species of identified pollen within 37 families. There was significant difference between locations and sampling period. The different geographical ranges determine various types of pollen.

DOI: 10.4018/978-1-6684-6265-2.ch010

PALYNOLOGY

Stingless bees are responsible for pollinating agents for many flowering plants and foraging pollen and nectar for food. Recently in Thailand, a melissopalynological analysis was carried out by examining 72 *Tetragonula pagdeni* honey samples obtained from different locations (Thakodee et al., 2018). This study aimed to identify the principal food sources (botanical origin) used by Sakagami et al., (1983) for honey production. The pollen grains were harvested from the honey samples and preserved in glycerin jelly and absolute ethanol for light microscopy (LM) and scanning electron microscopy (SEM) observation, respectively. In brief, 300 pollen grains per sample were counted, identified and compared with the pollen source catalogues of flowers. Morphological characteristics are symmetry, shape, polarity, apertural pattern, exine and ornamentation were determined. The study revealed that the rambutan (*Nephelium lappaceum*) was the dominant pollen type in all sampling locations. Then, pollen from foxtail palm, (*Wodyetia bifurcate*), coconut (*Cocos nucifera*) and sensitive plant (*Mimosa pudica*) are abundantly found in the honey samples. Meanwhile, pollen types of *Asystasia gangetica* (Acanthaceae), *Amaranthus lividus* (Amaranthaceae), *Areca catechu* (Arecaceae), *Chromolaena odorata* (Asteraceae) and *Durio zibethinus* (Malvaceae) were also found in the *T. pagdeni* honey.

On the other hand, the author elucidates that the stingless and honey bees are attracted to the aromatic scent of male and female rambutan's flowers. In general, stingless bees are known as major pollinators of palm trees, whereas palm trees sourced a good nectar for stingless bees. Thus, it was explained that the pollen from foxtail, coconut and *Areca catechu* in the *T. pagdeni* honey.

The stingless bees usually forage the food during day time. In this study, they found a smaller amount of durian (*D. zibethinus*) pollen in the *T. pagdeni* honey. However, this kind of tree are planted in everywhere around the experimental area (Thakodee et al., 2018). This is because, the durian flowers blooming at night which serve and attract the nocturnal pollinator such as bad and insects. Remarkably, during the dearth of the fruits or agricultural plants the stingless bees in general alter their food source and foraging the weeds. For example, *Asystasia gangetica* (Acanthaceae), *Amaranthus lividus* (Amaranthaceae), *Chromolaena odorata* and *Mikania cordata* (Asteraceae), *Mimosa pudica* (Fabaceae) and *Pennisetum pedicellarum* (Poaceae) pollens were identified in the *T. pagdeni* honey. Hence, this study revealed that mechanism in *T. pagdeni* maintaining their colonies during flowering and dearth seasons and also emphasized the importance of this species as great pollinators in the local plants. In addition, this study could be the diligently reference since it was carried out in Thailand where similar finding could be disclosed in other Southeast Asia country.

POLLEN IDENTIFICATION

The pollen sources of *H. itama* were identified by studying pollen grains present in pollen pots. Total of 66 pollen types was identified from pollen pots of *H. itama* collected from the hives in Jeli, Tanah Merah and Kota Bharu (Table 1). These pollens could be classified into 37 family plants based on the results. Pollen identification was shown in Figure 1 until Figure 9.

ANALYSIS OF TOTAL POLLEN FORAGED BY H. ITAMA

The *H. itama* foraged different types of pollen which could be divided the pollen frequency into three classes such as secondary pollen (16-45%), important minor pollen (3-15%) and minor pollen (1-3%) (Table 2). These pollens were grouped in respective classes based on the frequency present in recent study. The pollen study was carried out in a four seasons per year at three different locations and in this study, predominant pollen could not be classified since there is no dominant pollen frequency exceeds more than 45%. Based on the observation, only two types of pollen were found foraged by the *H. itama* in secondary pollen type class (16-45%). Then, six pollen types and 13 pollen types were identified at frequency 3-15% and 1-3% respectively.

Figure 1. Pollen types present in H. itama pollen pot.
Note: *1- Portulaca grandiflora, 2- Zea mays, 3- Nephelium lappaceum, 4- Cucumis sativus L., 5- Veitchia merillii, 6- Asystasia intrusa, 7- Cocos nucifera, 8- Nelumbo nucifer (Range of pollen size: 60-150μm)*

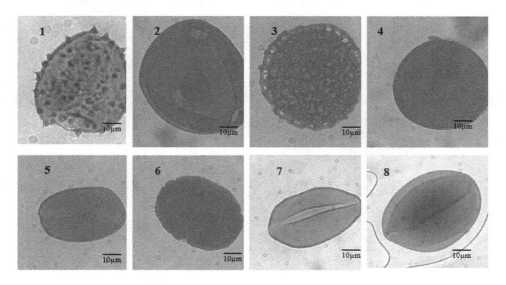

Figure 2. Pollen types present in H. itama pollen pot.
Note: 9- *Thevetia pruviana*, 10- *Ocimum basilicum L.*, 11- *Orthosiphon aristatus*, 12- *Musa acuminate*, 13-*Ipomoea pes-capre L.*, 14- *Leuceane leucocephala*, 15- *Antigonan leptopus*, 16- *Sesamum orientale* (Range of pollen size: 58-150μm)

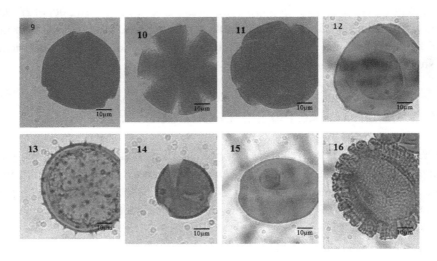

Figure 3. Pollen types present in H. itama pollen pot.
Note: 17- *Jacandra obtusifolia*, 18- *Phytchosperma macarthurii L.*, 19- *Nypa fruticans*, 20- *Acacia auriculiformis*, 21- *Ananas comosus*, 22- *Durio zibethinus*, 23- *Tecoma stans L.*, 24- *Acacia holosericeae* (Range of pollen size: 40-70μm)

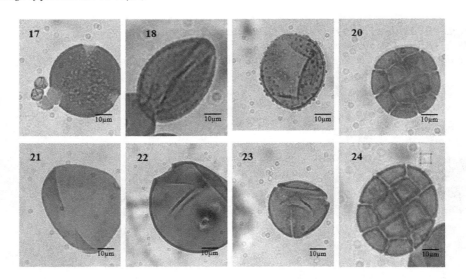

162

Figure 4. Pollen types present in H. itama pollen pot.
Note: 25- Eugenia malaccensis, 26- Cyperus brevifolius, 27-Rhizopora mucronata L., 28- Capsicum frutescens, 29- Avicennia alba, 30- Cleoma rutidosperma, 31- Cyperus kyllingia, 32-Mimosa separia (Range of pollen size: 20-28μm)

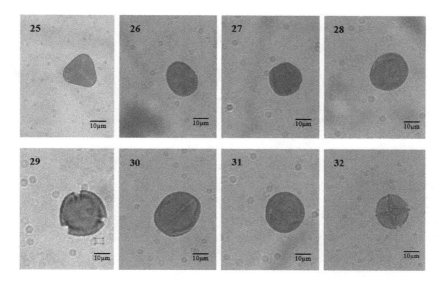

Figure 5. Pollen types present in H. itama pollen pot.
Note: 33- Tridax procumbens, 34- Sporobolos indicum, 35-Averrhoa carambola L., 36- Dillenia suffruticosa, 37- Pluchea indica L., 38- Solidago virgaurea L., 39- Melastoma malabathricum L, - 40- Coffea canephora (Range of pollen size: 20-30μm)

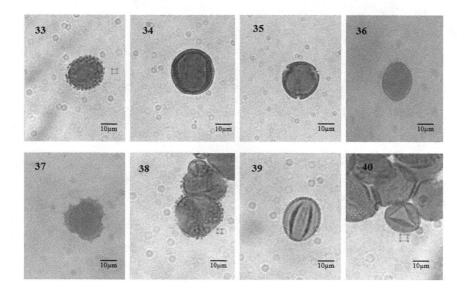

Figure 6. Pollen types present in H. itama pollen pot.
Note: *41- Hevea brasiliensis, 42- Gliricidia sepium, 43-Erythrina orientalis, 44- Bidens pilosa, 45-Vernonia cinerea, 46- Ishaemum muticum L., 47-Commelina diffusa, 48- Mimosa invisa (Range of pollen size: 30-40μm)*

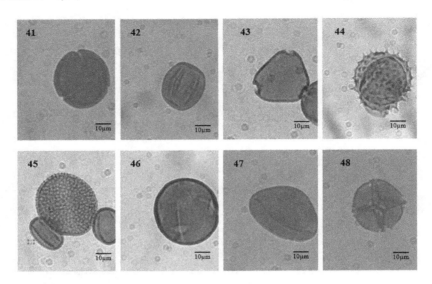

Figure 7. Pollen types present in H. itama pollen pot.
Note: *49- Cinnamomum verum, 50- Citrus aurantifolium, 51-Averrhoa bilimbi L., 52- Moringa pterygosperma, 53-Areca catechu, 54- Nerium indicum, 55- Wedelia biflora, 56- Mimusops elengi L. (Range of pollen size: 30-45μm)*

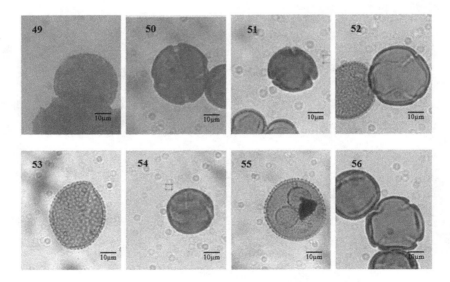

Figure 8. Pollen types present in H. itama pollen pot.
Note: 57- *Artocarpus heterophyllus*, 58- *Mimosa pudica*, 59-*Mikania cordata*, 60- *Muntigia calabura*, 61- *Bruguiera cylindrical L.*, 62- *Eupatorium odoratum L.*, 63- *Fagraea fragrans*, 64- *Callistemon speciosus (Range of pollen size: 10-21µm)*

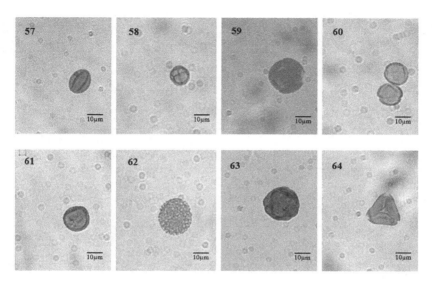

Figure 9. Pollen types present in H. itama pollen pot.
Note: 65- *Cassia biflora L.*, 66- *Vitex piñata (Range of pollen size: 30-40µm)*

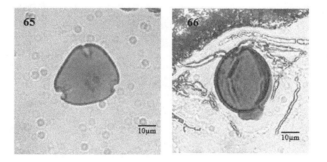

Table 3 shows total pollen foraged by *H. itama* in four period at three locations. The highest frequency pollen was significantly found during period four in Jeli. Among three locations, the frequency pollen in total was significantly abundant in Jeli compared with other locations. The *H. itama* actively foraged in period two where all locations exhibited in large frequency. Empirical data shows the *H. itama* in Tanah Merah farm foraged low types of pollen which represent by the pollen frequency such as in Table 2.

Table 1. List of pollen types found in the honey of H. itama

Scientific Name	Common Name	Local Name
Cyperus brevifolius	Short-leaved Cyperus	
Rhizopora mucronata Lam.	Mucronate mangrove	Bakau kurap
Capsicum frutescens	Chilli	Cili/cabai
Hevea brasiliensis	Rubber	Pokok getah
Zea mays	Maize	Pokok jagung
Gliricidia sepium	Mexican lilac tree	
Avicennia alba	White avicennia	Api-api putih
Nephelium lappaceum	Rambutan	Rambutan
Cleoma rutidosperma	Purple cleome	Mamam
Veitchia merillii	Manila palm	
Asystasia intrusa	Common asystasia	Pokok israel/
Thevetia peruviana	Yellow oleander	
Cucumis sativus L.	Cucucmber	Timun
Citrus aurantifolium	Lime	Limau nipis
Cyperus kyllingia Endl.	White Cyperus	
Leuceane leucocephala	Lead tree	Petai jawa
Tecoma stans L.	Yellow bell	Locing kuning
Portulaca grandiflora Hook		Ros jepun
Erythrina orientalis	Indian coral tree	Dedap
Ptychosperma macarthurii	Palmae	
Eugenia malaccensis	Malay apple	Jambu bol
Mikania cordota	Mile-a-minute plant	
Commelina diffusa	Common spiderwort	Rumput aur
Muntigia calabura	Malayan cherry tree	Buah ceri
Mimosa invisa	Scrambling sensitve plant	
Acacia auriculiformis	Acacia	Pokok akasia
Averrhoa carambola L.	Starfruit	Belimbing
Dillenia suffruticosa	Shrubby simpoh	Pokok simpoh air
Verronia cinerea L.	Common Vernonia	
Oluchea indica L.	Malayan flea bane	Beluntas
Moringa pterygosperma	Drum-stick tree	Keloh
Cinnamomum verum	Cinnamon	Kayu manis
Orthosiphon aristatus	Cats whiskers	Misai kucing
Mimosa pudica	Sensitive plant	Pokok semalu

continued on following page

Table 1. Continued

Scientific Name	Common Name	Local Name
Areca catechu	Betel-nut palm	/pinang
Antigonan leptopus	Honolulu creeper (coral vine)	Airmata pengantin
Tridax procumbens L.	Coat buttons	pokok kancing baju
Ipomoea pes-capre L.	Sea morning glory	Pokok tapak kuda
Gliricidia sepium	Mexican lilac tree	
Jacaranda obtusifolia	Jacandra	Pokok jambul merak
Mimosa sepiaria	Giant mimosa	
Cocos nucifera L.	Coconut	Kelapa
Musa acuminata	Banana	Pisang
Nerium indicum	Oleander	
Acacia holosericeae	Silver wattle	Akasia perak
Sesamum orientale	Rain tree	/pokok pukul lima
Bidens pilosa	Begger-tick	Kancing baju
Solidago virgaurea L.	Golden rod	
Artocarpus heterophyllus	Jackfruit	Pokok nangka
Melastoma malabathricum L.	Singapore rhododendron	Senduduk
Ananas comosus	Pineapple	Nenas
Coffea canephora	Coffee	Pokok kopi
Averrhoa bilimbi L.	Starfruit/carambola	Belimbing
Bruguiera cylindrical		Berus
Sporobolus indicum L.	Common dropseed	
Wedelia biflora	Sea oxeye	Serunai
Durio zibethinus	Durian	Durian
Mimusops elengi L.	Tanjung tree	Bunga tanjung
Eupatorium odoratum L.	Siam weed	Pokok kapal terbang
Ishaemum muticum L.	Seashore centipade plant	Rumput tembaga jantan
Nypa fruticans	Palmae	Nipah
Nelumbo nucifera	Lotus	Teratai
Fagraea fragrans		Tembusu
Callistemon speciosus	Bottle brush tree	Pokok berus botol
Cassia biflora L.	Bushy cassia	
Vitex piñata	Malayan teak	Pokok leban

Table 2. Pollen foraged by H. itama based on the pollen frequency for all locations.

Secondary Pollen Type (16-45%)	Important Minor Pollen Type (3-15%)	Minor Pollen Type (1-3%)
Cyperus kyllingia	Averrhoa carambola	Commelina diffusa
Eugenia malaccensis	Dillenia suffruticosa	Verronia cinerea
	Mimosa pudica	Oluchea indica
	Ishaemum muticum	Moringa pterygosperma
	Hevea brasiliensis	Antigonan leptopus
	Citrus aurantifolium	Cocos nucifera
		Nerium indicum
		Solidago virgaurea
		Artocarpus heterophyllus
		Ananas comosus
		Rhizopora mucronata
		Nephelium lappaceum
		Cleoma rutidosperma

Table 3. Total pollen foraged by H. itama collected from three locations in four periods.

Locations	Period	Total Pollen Collected
Jeli	September 2014	40.900±27.87
	December 2014	524.428±1325.31
	March 2015	178.385±194.42
	June 2015	700.875±1675.44
Tanah Merah	September 2014	44.083±88.67
	December 2014	70.000±123.41
	March 2015	47.944±96.32
	June 2015	24.191±42.29
Kota Bharu	September 2014	24.722±27.87
	December 2014	117.750±296.15
	March 2015	13.708±20.90
	June 2015	24.000±28.49

The analysis of pollen or palynology study currently revealed that the *H. itama* is an important pollinator in the ecosystem. A total number of 66 pollen types were identified from the collected pollen throughout the experiment (Table 2) which anticipated to be similar in another area in Kelantan. As a comparison, the previous study revealed that *H. itama* in coastal areas foraging pollen grains from plants such as *Antigonan leptopus, Amaranthus tricolor, Hibiscus rosa-sinensis, Cucumis melo, Ixora coccinea, Tridax procumbens, Biden pilosa, Turnera subulata* and *Ixora javanica* (Lob et al., 2017). The author also highlighted that the *Portula grandiflora* is predominantly found foraged by the *H. itama* (Lob et al., 2017). Meanwhile, in recent study areas, the *Cyperus kyllingia* and *Eugenia malaccensis* are dominantly found at 24 to 29% respectively. *Cyperus kylingia* pollen was found foraged by *H. itama* in all study locations whereas *E. malaccensis* abundantly found in the pollen pot of *H. itama* only in Jeli during June 2015. This happened due to the study was carried out during the flowering season of *E. malaccensis* in the Jeli area which solely contributed to the foraging activity of *H. itama*. Thus, this occurrence has disclosed the significant roles of the *H. itama* as a pollinator agent for the tropical forest, particularly for this plant. The roles played by the stingless bees are equivalent to the bees where they are significant in plant reproduction, contribute to plant diversity, improving quantity and quality of fruits and vegetables (Gallai et al., 2009; Zerdani et al., 2011). Therefore, the abundance of the pollen grains in the pollen pot may anticipate the production of the fruits season and production as well.

On the other hand, the important minor pollen type was found in six different plant families. These pollens such as *Averrhoa carambola, Dillenia suffruticosa, Mimosa pudica, Ishaemum muticum, Hevea brasiliensis*, and *Citrus aurantifolium* were found foraged by the *H. itama*. This finding has revealed that the *H. itama* specifically foraged the pollen grains from specific plant groups. Zerdani et al. (2011) coined that the specific pollen grains were collected by the specific bees. In this case, *H. itama* seems to be important pollinators since it has a wide range of pollen foraging capacity from shrubs, herbs, grass, vegetables, and tree. These plants produce flowers for almost throughout the year although in drought season where these plants require little maintenances and serve as *H. itama* food sources. However, during the monsoon period, the frequency of the pollen grains in the samples was found to be little compared to the other period. Probably, the short period of foraging activity could be done and little availability of flowers during this period. Furthermore, the foraging activity of the *H. itama* changes from season to season. For instance, previous research coined that the stingless bees change their foraging activity from economic plants to weeds during the dearth period (Thakodee et al., 2018). Thus, although the pollen weeds are not as significant as other economic plants it is useful for the meliponiculture to maintain the colonies in the dearth period.

Besides, the capability of the *H. itama* in carrying the pollen grains are contingent to the pollen sizes. In recent findings, large pollen grain size also found in the sample such as *Durio zibenthinus*. This plant grain only found once in the sample of pollen from Kota Bharu. Surprisingly, this plant particularly flowering at night and another large pollinator are responsible to take the pollen grains but recent observation may suggest the large *H. itama* worker took the pollen from the fallen flowers. A similar case in *Acacia holosericeae* and *Sesamum orientale* only one pollen grain was found in the Jeli sample. These plants probably have similar pollen characteristics such as in *D. zibenthinus* or the low existence of the flowers.

The distribution of the pollen grains also correlated to the flowering characteristics of the plants. For example, current finding shows the pollen foraged such as in-class important minor and minor types could be found at all period of study in all locations. The previous studies highlighted the foraging activity of the *H. itama* highly contingent to pollen preference such as shape, attractiveness, and availability (Bahri, 2018). Gilman (1999) coined the attractiveness of *Ixora coccinea* flowers that existed in orange, yellow, pink, bright red or white colors and bloom continuously has contributed to high frequency in pollen grains foraged by the *H. itama*. On the other hand, the pollinator such as *H. itama* also attracted to the strong and nice odor (Bahri, 2018). For instance, in a recent study found the *A. carambola* has a high frequency of the pollen grains forage by *H. itama* in all periods. Apparently, this plant has attractants that entice the *H. itama* workers to forage the pollen. In addition, Azuma et al. (2002) found that the flower scent is a significant factor in attracting pollinator such as stingless bee species to forage. This attractant signalling the pollinators such as *H. itama* to recognize the flower sources. For example, the pollen grains from *N. lappaceum* or rambutan were found in evaluated samples particularly abundant in Jeli. This study revealed a high production of this fruit in Jeli compared with other evaluated areas. On top of that, this plant highly sourced food to *H. itama* in this area. A similar finding was found in Thailand where rambutan as a native crop serves food for commercial *Tetragonula pagdeni* in eastern area (Thakodee et al., 2018). The authors highlighted the *T. pagdeni* assisting the pollination of rambutan from December to January. Meanwhile, the recent finding indicates the pollination actively from June to August and continues until early December. The *H. itama* has a preference for foraging the rambutan nectar where the aromatic male and female flowers attract them or honey bees (Shivaramu et al., 2012). Hence, appropriately to say that the H. itama as important insect species to produce rambutan.

On the other hand, *H. itama* also foraging the rubber flower in a recent study which classified as an important minor pollen type due to its frequency found in all locations at four periods. Previous research by Bahri, (2018) was revealed contradict finding where the *H. itama* does not forage the rubber flowers. The author mentioned the *H. itama* could not forage due to the height of the plant which causes difficulties to the bee workers to forage the pollen. Therefore, the recent findings may disclose the paradox of the capability of the *H. itama* to forage pollen from the tall trees. In addition, there is an argument to agree that the *H. itama* likely to forage the nectar from the extrafloral nectaries of *H. brasiliensis*. Recent finding elucidates this paradox by disclosing an example of pollen grains from the *C.nucifera*. This tree marginally tall than rubber tree but *H. itama* could forage the coconut pollen throughout the year. This finding agreed with the previous study where the *T. pagdeni* predominantly foraging this pollen throughout the year as well in the eastern part of Thailand where the coconut abundantly being planted commercially. However, this plant pollen was found none in Jeli which indicates no trees were planted near to the hives. Furthermore, this minor pollen type in recent analysis revealed that *H. itama* is not a good pollinator for this plant compared to the honey bees, Apis spp. Therefore, the height of the plants is not the main factor of the pollen absence in the palynology study and different species of pollinator may have different capabilities to forage the plants' pollen.

The analysis of pollen in a recent study has revealed that the pollen grains forage by the *H. itama* has significant differences between locations and period. For example, the highest pollen frequency was found in the Jeli at June 2015. This finding was contributed by the flowering seasons and the availability of the food sources that attracting the *H. itama* to forage. Although the pollen frequency was higher in the Jeli area at all periods, however the types of pollen variety abundantly found in Kota Bharu. This finding may explain the geographical area and distribution of the plants. For instance, 66 types of pollen grains were identified in the Kota Bharu sample due to this area particularly build for stingless bee farming and many plant species were planted for their food source. In comparison, only 35 types of pollen were identified from the Jeli sample which disclosed that the pollen grains foraged from the forest plants such as *Cinnamomum verum*. This pollen grain only found in Jeli instead of pollen grains from *Gliricidia sepium, Thevetia peruviana, Portulaca grandiflora, Erythrima orientalis, Eugenia malaccensis, Mikania cordota, Oluchea indica, Orthosiphon aristatus* and *Areca catecna*. The distribution of the plants plays an important role in the palynology of the *H. itama*. *Caffea canephora* or coffee plant was found being planted in Tanah Merah where its pollen was found in the sample. Meanwhile, the highest frequency of *E. mallacensis* in Jeli from December to February indicates the fruit season which significantly as an economic crop to local people. This plant was found abundantly distributed in the Jeli area but none in Kota Bharu and Tanah Merah.

CONCLUSION

A palynology study in the recent study has revealed the pollen grains that foraged from the plants. The different geographical range determines various types of pollen and their dominance. In addition, identification of pollen grains indicates the suitable of the *H. itama* as pollinator. Furthermore, the abundance of the pollen grains indicates the fruit season and their production. There were 66 types of pollen being identified from there farm. It was classified into 37 plant's family. Total of all pollen counted were 17097. Based on the pollen identification and pollen distribution can help the farmers to plant a suitable plant for meliponiculture.

ACKNOWLEDGMENT

We would like to thank the reviewers for all their constructive comments and suggestions.

REFERENCES

Azuma, H., Toyota, M., Asakawa, Y., Takaso, T., & Tobe, H. (2002). Floral scent chemistry of mangrove plants. *Journal of Plant Research*, *115*(1), 47–53. doi:10.1007102650200007 PMID:12884048

Bahri, S. (2018). Pollen profile by stingless bee (Heterotrigona itama) reared in rubber smallholding environment at Tepoh, Terengganu. *Malaysian Journal of Microscopy*, *14*(1).

Gallai, N., Salles, J.-M., Settele, J., & Vaissière, B. E. (2009). Economic valuation of the vulnerability of world agriculture confronted with pollinator decline. *Ecological Economics*, *68*(3), 810–821. doi:10.1016/j.ecolecon.2008.06.014

Gilman, E. F. (1999). Ixora coccinea. *University of Florida, Cooperative Extension Service. Institute of Food and Agriculture Sciences. Fact Sheet, FPS-291*, 1–3.

Lob, S., Afiffi, N., Razak, S. B. A., Ibrahim, N. F., & Nawi, I. H. M. (2017). Composition and identification of pollen collected by stingless bee (Heterotrigona itama) in forested and coastal area of Terengganu. *Malaysian Applied Biology Journal*, *46*(3), 227–232.

Sakagami, S. F., Yamane, S., & Inoue, T. (1983). Oviposition behavior of two Southeast Asian stingless bees, Trigona (Tetragonula) laeviceps and T.(T.) pagdeni. 昆蟲, *51*(3), 441–457.

Shivaramu, K., Sakthivel, T., & Reddy, P. V. (2012). Diversity and foraging dynamics of insect pollinators on rambutan (Nephelium lappacum L.). *Pest Management in Horticultural Ecosystems*, *18*(2), 158–160.

Thakodee, T., Deowanish, S., & Duangmal, K. (2018). Melissopalynological analysis of stingless bee (Tetragonula pagdeni) honey in Eastern Thailand. *Journal of Asia-Pacific Entomology*, *21*(2), 620–630. doi:10.1016/j.aspen.2018.04.003

Zerdani, I., Abouda, Z., Kalalou, I., Faid, M., & Ahami, M. (2011). The Antibacterial Activity of Moroccan Bee Bread and Bee-Pollen (Fresh and Dried) against Pathogenic Bacteria. Res. *Journal of Microbiology (Seoul, Korea)*, *6*, 376–384.

Chapter 11

Phenolic and Flavonoid Content of Propolis Extracts of *Heterotrigona itama* From Rubber Smallholding Area and Forestry Surrounding Area

Nora'aini Ali
Faculty of Ocean Engineering Technology and Informatics, Universiti Malaysia Terengganu, Malaysia

Norafiza Awang
Faculty of Ocean Engineering Technology and Informatics, Universiti Malaysia Terengganu, Malaysia

Norhafiza Ilyana Yatim
Centre of Lipids Engineering and Applied Research (CLEAR), Ibnu Sina Institute of Scientific and Industrial Research, Universiti Malaysia Terengganu, Malaysia

Norasikin Othman
Higher Institution Centre of Excellence (HICoE), Institute of Tropical Aquaculture and Fisheries, Universiti Teknologi Malaysia, Malaysia

Shamsul Bahri Abd Razak
Apis and Meliponine Special Interest Group, Faculty of Fishery and Food Sciences, Universiti Malaysia Terengganu, Malaysia

ABSTRACT

The various botanical origins may be influenced by the type of plant used as a food source, which affects the chemical composition of propolis. The purpose of this

DOI: 10.4018/978-1-6684-6265-2.ch011

work was to determine the antioxidant activity, total phenolic content (TPC), and total flavonoid content (TFC) of propolis extracted from Indo-Malayan stingless bees, Heterotrigona itama, rearing at different botanical regions. Propolis was obtained from two different botanical origins: Forested area (H. Itama-FA) propolis from Taman Pertanian Sekayu, Terengganu and Hevea brasiliensis (HB) propolis from stingless bees that reared in the rubber smallholding at Bukit Berangan, Terengganu (H. Itama-HB). TPC and TFC concentrations were evaluated using a UV-Vis Spectrophotometer, whereas antioxidant activity was determined using the DPPH free radical assay method. The results showed that the propolis of stingless bees rearing in rubber smallholdings area and the wildly available in forest area have comparable quality in terms of promising sources of antioxidant compounds.

INTRODUCTION

The Indo-Malayan clade has more than 50 species of stingless bees. The most often domesticated and bred stingless bee species include Heterotrigona itama (H.Itama), Geniotrigona thoracica, Lepidotrigona terminate, and Tetragonula laeviceps (Slaa et al., 2006). In Malaysia, forests are the most popular domesticated areas for stingless bees, which contains various tree species that could support great biodiversity for the stingless bees colony production and survival. A heterogeneous culture in rubber plantations, such as mixing rubber trees with other plants or animals. A heterogeneous culture is one of the ongoing efforts to ensure the sustainability of rubber industry sectors and the socioeconomic well-being of smallholder farmers whose livelihoods are heavily reliant on rubber tapping. The farmers face significant obstacles due to the inherent vulnerability of rubber prices to global economic downturns, despite the numerous limits imposed by climate change and environmental concerns. There is a downward trend in latex pricing due to the decline in demand for the rubber industry. As a result, numerous initiatives and approaches have been made to integrate diverse activities with rubber plantations to provide additional revenue for farmers.

Since the early 1980s, the Rubber Research Institute of Malaysia (RRIM) has been integrating meliponiculture, such as stingless beekeeping, into rubber growing areas, mostly in rubber smallholder areas. Previous research and pilot programmes have demonstrated the feasibility and viability of growing stingless bees in rubber tree habitats (Rao & Vijayakumar, 1992). Thus, integrating stingless bees, particularly Heterotrigona itama, into rubber plantations might be considered a secondary source of revenue for rubber smallholders while simultaneously serving as an important pollination agent for the crop (Zaki & Razak, 2018). However, scientific facts and full knowledge of the Indo-Malayan stingless bees are still lacking in Malaysia. As a result, farmers have not yet benefited from successful information transfer about stingless beekeeping and rearing.

Natural rubber is primarily produced by the rubber tree (Hevea brasiliensis). It is perennial and has a 30-35-year economic life. Adult trees display a distinct annual flushing pattern known as wintering, during which they shed their leaves almost completely for 3–4 weeks before commencing fresh flushes. Hevea flowering, which occurs once a year on average, begins after leaf fall and is strongly influenced by climatic variables. Malaysia's tropical rainforest climate, with annual rainfall ranging from 2000–2500 mm and an average temperature of 26–28°C, is ideal for commercial rubber planting. Malaysia's dry and wet seasons were highly correlated with H. itama-HB foraging activities. Due to their inability to fly on wet days, foraging activity slows. According to a prior study, the optimal harvesting period for honey stored under rubber trees was between February and March, during the rubber trees' flowering season (Tajuddin, 1986). Preliminary findings indicate that a single colony can produce an average of 3 kg of honey per flowering season (Sujan et al., 1984).

Apart from honey, the primary product of stingless beekeeping, propolis and beebread are significant by-products. Propolis is a resinous compound of resin, leaf buds, mucilage, gum, and beeswax. The majority of stingless bees generate propolis similarly to honey bees. As a result, the qualities and properties of propolis are highly dependent on the stingless bees' departures and returns as foragers and the type of materials they transported. Prior nutraceutical research has focused on honey bee propolis rather than stingless bee propolis. Due to the chemical components found in propolis, it could be employed as a basic ingredient in pharmaceutical and cosmetic goods. Numerous beneficial chemicals, including flavonoids and phenolic compounds, can be isolated from propolis (Awang et al., 2018). The nutritionist emphasised the medicinal qualities of dietary flavonoid and phenolic compounds as bioactive substances with various therapeutic applications in human health (Panche et al., 2016). While propolis is a valuable by-product produced by stingless bees, it is not commercially exploited due to local beekeepers' lack of expertise, procedures, and even awareness. Thus, studying the untapped propolis of stingless bees is critical for further exploiting competitive advantage opportunities and additional values.

As previously stated, the various botanical sources play a significant role in determining the sorts of plants foraged by stingless bees, which are the primary source of elements that impact the chemical components of propolis. These compositions vary according to the vegetation type in the area from where they were taken (Castaldo & Capasso, 2002; Kumazawa et al., 2004). According to their geographical or botanical origins, propolis from Europe, South America, and Asia will have significantly different chemical compositions (Marcucci, 1995). Propolis' chemical composition can also change according to the season, the location where the plant resin is collected, and the bee species (Castro et al., 2007; Mercan, 2006). Thus, rubber tree propolis is expected to possess the unique properties of propolis chemical compounds, including rubber latex, which we consider is at least on par with propolis production in other origins. This study aims to determine the antioxidant activity, total phenolic content, and total flavonoid content of propolis generated by the Indo-Malayan stingless bee, Heterotrigona itama, from two distinct botanical sources: forest area and rubber smallholder. The findings of this study may be utilised to influence the selection of an appropriate landscape for meliponiculture, particularly in the rubber environment (Zaki & Razak, 2018). Additionally, the findings may offer value to stingless bee rearing in rubber plantations. Even though meliponiculture in rubber plantations is a relatively new endeavour in Malaysia, it generates considerable attention and is gradually adopted by the people. This activity is fast gaining interest and consideration as a potential source of revenue or possibly a secondary source of income for rubber smallholders (Abd Razak et al., 2016).

METHODOLOGY

Propolis Samples Collection and Preparation

In this study, we investigated propolis from stingless bees, Heterotrigona itama (H. itama), as the propolis is abundantly available and could be collected significantly. We intend to compare the availability of Total Phenolic Content and Total Flavanoid Content from stingless bees from two distinct environmental characteristics; (a) were wildly available in the forest area of Taman Pertanian Sekayu, Hulu Terengganu, and (b) were purposefully reared in rubber smallholding area in Bukit Berangan, Tepoh, Terengganu.

Chemicals and Reagents

Aluminium chloride (AlCl3) and all ethanol, methanol and acetone were purchased from Merck, Germany. All reagents of 1,1-diphenyl-2-picrylhydrazyl (DPPH) and Folin Ciocalteu's were acquired from Sigma-Aldrich, USA. All chemicals and reagents utilised were of analytical grade quality without further purification.

Preparation of Extract

Methanol was the most effective solvent for extracting bioactive antioxidant compounds. To begin, samples of Heterotrigona itama (H. itama) stingless bees were collected from raw propolis and then frozen at a temperature of -10 °C. The samples were then ground into powder to improve the surface area, and all the samples were stored and desiccated until the extraction began. The methanolic extraction method extracted the raw propolis of stingless bees with slight modifications (Abd Jamil & Zhari, 2016). The raw propolis sample was crushed and macerated overnight in acetone at room temperature. The filtrate was obtained by gently percolating the material, and the leftover sample was soaked in 70% methanol. This procedure was continued until the filtrate lost its colour. The filtrates were mixed, and the methanol was extracted using a rotary evaporator set to 45°C to obtain the dry methanolic extract of propolis. Before usage, the extract was kept at 4°C.

The Yield of Extraction

The yield of extraction is calculated using Equation (1) below:

$$Yield = \left[\frac{Weight\,of\,Propolis(g)}{Weight\,of\,Raw\,Propolis(g)} \right] \times 100 \tag{1}$$

Preparation of Standard Solution

10 mg of gallic acid and 10 mg of quercetin were carefully weighed into a 10 ml volumetric flask to prepare a solution of concentration 1 mg/mL. Methanol was used as a standard solution to dissolve both compounds.

Determination of Total Phenolic Content

The total phenolic content was determined using a modified Folin-Ciocalteu colorimetric technique (Cicco & Lattanzio, 2011; Oliveira et al., 2016). This approach is based on the oxidation and reduction of the phenolate ion in alkaline conditions. Follin Ciocalteu's $MO6+$ and $W+$ complex ions decrease, changing the blue colour reaction (Medic-Saric et al., 2013). Gallic acid was utilised as a standard, and total phenolics were expressed as mg gallic acid equivalents per g of sample extract (mg GAE/ g extract). Gallic acid solution was prepared in ethanol as a standard solution at concentrations 1.0, 1.95, 3.9, 7.8, 15.63, 31.25 and 62.5 mg/ml. Propolis extract was also prepared in ethanol at a concentration ratio of 1:10 (0.1). Then, 0.5 ml of each sample was added to the test tubes and mixed with 0.5 ml of the Folin-Ciocalteou reagent and 0.5 ml of 7.5% sodium carbonate, $Na2CO3$ solution. After 2 h of incubation in the dark at room temperature, the absorbance at 760 nm was determined using a UV-Vis spectrophotometer. The value of gallic acid was determined using the standard calibration curve. All the procedure was repeated triplicate and expressed as average $\pm$ standard error of the mean.

The Procedure to Determine the Total Flavonoids Content (TFC)

Flavanoids were determined using the aluminium chloride colorimetric technique (Cunha et al., 2004; Kumazawa et al., 2004). The method employed quercetin as a reference, and flavonoid concentrations were determined using the quercetin equivalent (QE) per g of sample extract. 1 mL of the standard solution of concentration 1.95, 3.9, 7.8, 15.63, 31.25, 62.5 and 125 mg/mL of quercetin were prepared in ethanol. 0.5 mL of each sample (1 mg/mL) was mixed with 0.5 mL of 2% aluminium chloride, AlCl3 solution in a test tube. With a sample-to-solvent ratio of 1:10, the highest percentage of extracts was obtained (Mokhtar, 2019). After 1 h in the dark at room temperature, the concentration of remaining quercetin was determined using the UV-Vis spectrophotometer at 420 nm absorbance and calculated based on the calibration curve. All the procedure was repeated triplicated and expressed as average $\pm$ standard error of the mean.

The Radical Scavenging Activity of DPPH

The free radical scavenging activity DPPH was determined using the modified methods of (Blois, 1958) and Jo et al. (2012). The DPPH solution (0.2mM) was prepared in methanol and 100μl to 100μl of the sample extracts, 5.0, 2.5, 1.25, 0.6, 0.3, 0.16, 0.08, 0.04, 0.02, 0.01, 0.005 (mg/ml). The extracted solutions were stored at room temperature in a dark place for 30 minutes, and determined the absorbance at 515 nm wavelength. The radical scavenging activity was reported as an inhibition percentage using the following formula.

$$\% \, Radical \, Scavenging \, Activity = \left[\frac{Abs \, Control - Abs \, Sample}{Abs \, Control} \right] \times 100 \qquad (2)$$

RESULTS AND DISCUSSION

Our previous research has studied the antioxidant activity of two commonly sticky propolis species, H. itama and G. thoracica, as well as hard propolis species, H. aliceae, H. fimbriata, T. apicalis, T. vidua, T. peninsularis, L. canifrons, T. melanoleuca, and T. binghami. We confirmed that the propolis extract from H.itama species contains the highest level of Total Flavonoid Content (TFC) and Total Phenolic Compound (TPC), as well as the strongest antioxidant activity, when compared to other types of species (Awang et al., 2018). Honey is the primary product of meliponiculture. While other by-products like as propolis and bee bread (fermented pollen) may command a greater retail price than honey, these two by-products have little commercial use. Thus, this initiative attempts to demonstrate the benefits of cultivating H.itama stingless bees in rubber smallholder areas, emphasising the value of propolis's bioactive chemicals. We compared the TFC, TPC, and antioxidant activity of propolis extracts from H. Itama stingless bees reared in rubber smallholdings and stingless bees abundant in the adjacent forest.

Effects of Two Different Botanical Origins to the TFC, TPC and Antioxidant Activity of the Propolis Extract

H.itama species have been chosen and reared at small rubber holding to compare the bioactive compounds and antioxidant activities closely related to the surrounding environment or plant origins. We have analysed TPC, TFC and IC5 as research parameters to represent valuable bioactive compounds from the propolis extract of stingless bees. The rubber tree in the smallholding area where we reared the H. Itama stingless bees is matured and tapped. Initial results have shown that the honey yield from stingless bees in this vegetation area is seasonal. The peak for honey production is from January to April and starts to decline from April onward (Abd Razak et al., 2016). Thus, we have analysed the propolis samples of H.itama at forest area (H. itama-FA), and H. itama at rubber smallholding (H. itama-HB) collected in February due to the flowering period of rubber trees. Table 1 and Figure 1 show the average values of TPC, TFC and IC50 between two different geographical origins of propolis H.itama-FA and H. itama-HB. The results showed that the TPC for both H.itama-FA and H.itama-HB are 48.6 ± 1.50 mg/ml and 58.6 ± 1.31 mg/l, about 17% of the percentage difference. While, for the TFC results, H.itama-FA was 40% higher than H.itama-HB, which were 33.8 ± 9.07 mg/ml and 20.01 ± 1.31 mg/ml, respectively. The results of antioxidant activity are expressed as IC50. The lower IC50 value indicates the stronger antioxidant activity in propolis. Antioxidant activity for H. itama-FA was 2.6 times higher than H. itama-HB, which were 280 and 720 g/ml, respectively. This finding shows that the propolis of stingless bees rearing in rubber smallholdings shows about half the quality of the wildly available in the forest but could still be considered promising sources of antioxidant compounds.

The modest variations in TFC, TPC, and antioxidant activity values could be explained by the diverse geographical origins of the plant varieties used to make propolis. Thus, the chemical contents of propolis vary according to the plants from which it is derived. Propolis has a broad range of bioactivity, which is attributed mostly to the habitat of bee species and their food source, which varies by area. The composition of propolis varies according to the bees' feeding preferences, resulting in varying antioxidant capabilities. Although relatively small, the differences in

Table 1. The TFC, TPC, IC50 and Percentage difference of two different botanical origins from

Parameters	Botanical Origins		Percentage Difference (%)
	H.itama-**FA***	*H.itama*-**HB***	
TFC (mg/ml)	33.82 ± 9.07	20.01 ± 0.86	19.5 19.5
TPC (mg/ml)	58.6 ± 1.31	48.6 ± 1.50	0.8 0.8
IC$_{50}$ (mg/ml)	280	720	3.7

Note: *Values represent the mean of three determinations ± the standard deviations and average values.

Figure 1. Graph of the TFC, TPC and IC50 of propolis extract of two different botanical origins

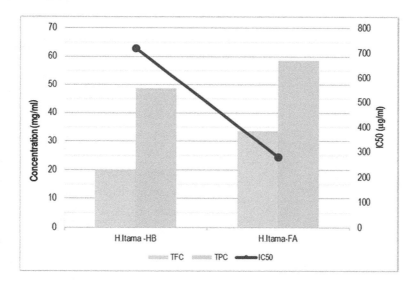

these values show that the pollen study could portray the plant's geographical origin, floral sources, and genus that the stingless bees visited (Ponnuchamy et al., 2014).

In this context, each flower species of floral source has a unique pollen grain due to the honey's geographical origin and major floral sources. In addition, the pollination efforts of stingless bees, either domesticated or wild colonies, are their most important characteristics, not only from the agricultural aspect but also in the natural environment. Approximately one-third of tropical rainforest flora depends on social bees' pollinating efforts (Oldroyd & Wongsiri, 2009). Thus, preserving and multiplying those foraged plants by stingless bees is crucial. Table 2 shows the previous studies about the types of pollen based on plant species foraged by H.itama

Table 2. Types of pollen-based on plant species foraged by H.itama-FA and H.itama-HB

Botanical Origins	
Taman Pertanian Sekayu, Terengganu (Forest Area) (Suhaizan et al 2017)	Bukit Berangan, Tepoh, Terengganu (Rubber smallholding) (Zaki & Razak 2018)
Bunga air mata pengantin (*Antigonan leptopus*)	Bunga kemboja (*Adenium obesum*)
Bayam (*Amaranthus tricolor L.*)	Rumput tahi ayam (*Ageratum conyzoides*)
Rumput Israel (*Asytasia gangetica*)	Sambilatana (*Andrographis paniculata*)
Bunga beiden (*Biden pilosa*)	Rumput Israel (*Asytasia gigantic*)
Bunga Ruelia (*Reullia brittonia*)	Bunga beiden (*Biden pilosa*)
Bunga roselle (*Hibiscus sabdariffa L.*)	Bunga kosmos kuning (*Caudatus sulhureus*)
Bunga ros jepun (*Portulaca grandiflora*)	Kaliandra merah(*Calliandra haematocephala*)
Pollen A	Chinese honeysucker (*Combretum inducum*)
Pollen B	Rumput mutiara (*Hedyotis corymbosa*)
Pollen C	Bunga Jarum (*Ixora coccinea*)
Pollen D	Keembong merah (*Impatiens balsamina*)
Pollen E	Bunga melur (*Jasminum polyanthum*)
Pollen F	Bunga lantana (*Lantana camara*)
Pollen G	Senduduk (*Melastoma malabathricum*)
Pollen H	Gelam (*Melaleuca cajuputi*)
	Pokok semalu (*Mimosa pudica*)
	Bunga kemunting (*Rhodomyrtus tomentosa*)
	Kembang pagi (*Turnera subulata*)
	Pollen A
	Pollen B
	Pollen C
	Pollen D
	Pollen E

that contribute to the different TFC, TPC and antioxidant activity in forest areas and rubber smallholding.

Notably, H. itama-HB does not graze on rubber blossoms, as no pollen was found during the experiment. It could be due to the plant's height, making pollen foraging difficult for stingless bees. The stingless bees most likely collected nectar exclusively from Hevea brasiliensis extrafloral nectaries. Stingless bees choose flowers with specific features, such as size, colour, and fragrance. A little flowering plant such as Ixora coccinea was advised for planting (Zaki & Razak, 2018).

Effects of Dry, Inter-monsoon and Wet Seasons on the TFC, TPC and Antioxidant Activity of the Propolis Extract

Foraging activities in stingless bees as social insects are mostly determined by environmental variables such as food availability and time (Suhaizan et al., 2017). According to Biesmeijer & Ermers (1999), external factors such as environmental and colony conditions influence the level of stimulus exposure and the threshold response to foraging stimuli (Biesmeijer & Ermers, 1999). Naturally, foragers of stingless bees will collect nectar and pollen in response to the environmental conditions, such as accessible food storage and resources in the field. Numerous studies have revealed that meteorological conditions, light intensity, humidity, food availability, competition, and colony state affect honey bees and stingless bees foraging behaviours (Nascimento & Nascimento, 2012).

In the aspect of rubber (Hevea brasiliensis), previous studies have concluded that climate factors significantly impact rubber yield, mainly associated with latex production. For example, temperature and relative humidity will influence tree stomata regulation, eventually affecting the latex flow, while rainfall intensity affects smallholders tapping activity (Priyadarshan, 2011). Fadzli et al. (2022) studied 37 Hevea clones in different environmental conditions and found that an increase of 1oC minimum temperature was associated with a reduction in latex yields of 3 gt-1 t-1 (Ali et al., 2020). Therefore, our next objective is to understand how the rainfall pattern could affect the quality of bioactive compounds in propolis extract with the latex yield in dry, inter-monsoon and wet seasons. For this purpose, we have collected propolis extracts of H. itama-HB each month throughout the year.

Hevea branselliensis (HB), a tropical rainforest type of plant, thrives in high humidity and coexists well with various kinds of wild trees, climbing plants, and undergrowth. The average annual rainfall in a typical Peninsular Malaysia climate is approximately 2000-2500 mm, directly affecting latex yielding (Biesmeijer & Ermers, 1999). Seasons are classified into three types (Biesmeijer & Ermers, 1999) according to their rainfall patterns, as shown in Table 3. Table 4 summarises the average TPC, TFC, and IC50 values obtained from a monthly collection of stingless bee propolis extract throughout the year.

Table 3. Classification of the season for a yield of latex production based on typical rainfall patterns for Peninsular Malaysia

Month	Season	Yield of Latex Production
January to April	Dry	Low yielding period
May to August	Inter-monsoon	Medium yielding period
September to December	Wet	High yielding period

Table 4. Average values of TPC, TFC and IC50 of monthly propolis extract of stingless bees throughout the year.

Month	TFC (mg/ml)	TPC (mg/ml)	IC_{50} (µg/ml)
January	22.173±1.12	56.49±4.35	680
February	20.06±0.87	48.64±1.50	720
March	39.53±1.32	62.30±6.07	2480
April	24.42±1.07	66.83±1.82	2300
May	31.03±1.37	68.44±2.53	600
June	27.90±1.01	51.76±1.52	960
July	109.78±2.72	52.90±4.59	390
August	159.78±12.15	69.25±1.21	540
September	19.38±0.88	54.32±4.11	840
October	64.36±1.04	65.41±2.71	550
November	42.59±1.06	62.67±1.53	1220
December	26.42±5.48	72.95±3.25	560
AVERAGE VAUES	48.95 mg/ml	60.96 mg/ml	987 (µg/ml)

Note: *All data is triplicates and reported as an average value± std deviation.

As indicated in Table 3, this data is divided into dry, inter-monsoon and wet seasons. The TPC compounds of propolis extract are almost the same quantity, regardless of yielding latex season throughout the year, with an average value of 61 mg/l. However, a distinct high distribution of TPC was observed during the inter-monsoon season. The concentration of TPCs' gradually increased from May to August, whereby the highest peak was found in August, three times higher than the average year collection (110 mg/l). The second highest of TFC compounds is 110 mg/l, collected in July, about double the capacity of the average value. By the end of the inter-monsoon season, the TPC's concentration showed drastic drops as the wet season started and started to show a fluctuating pattern. These findings also reflect the fluctuation of propolis composition, which is directly related to the bees' foraging activities and preference for food resources. From our point of view, there was a strong correlation between the propolis characteristics and properties and rubber trees' growth mechanism as the H.Itama stingless bees were domesticated in rubber smallholdings areas. Rubber trees defoliated their leaves during the wet season (September to December) as a defence mechanism against water loss. During this time-period, the tree's leaves expire and fall off, allowing for the formation of new leaves. The sixteen-week wet season period significantly affects the tree's metabolism, latex constitution and yield. It explains the fluctuation trend of TPC value, which is an indicator of propolis quality during the wet season.

Figure 2. A monthly collection of propolis extract of stingless bees throughout the year: Inter-relation of the average values of TFC, TPC and IC50 of H. itama-HB propolis extract

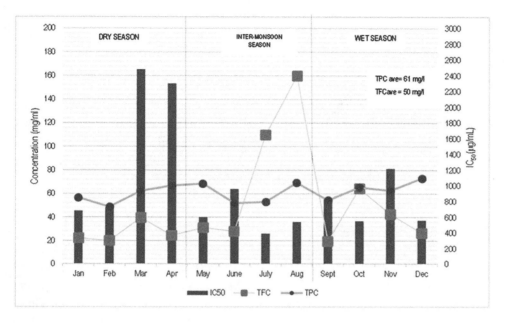

Figure 2 tabulates the amount of TPC and TFC compounds of propolis extract collected throughout the year. The inter-relation of three main variables, namely TPC, TFC and IC50, reflect propolis extracts' potential as precursors for functional food products. Together with phenolic compounds, flavonoids have various therapeutic properties for human health, including antioxidant, anti-inflammatory, anticancer and antimicrobial activities (Al-Hatamleh et al., 2020). The IC50 value is a parameter widely used to measure the antioxidant activity of test samples. It is used to measure the potency of TPC and TFC in inhibiting a specific biological or biochemical function. Thus, the lower the value of IC50, the higher the antioxidant activity is. IC50 is calculated as the concentration of antioxidants needed to decrease the initial DPPH concentration by 50% (Sánchez-Moreno et al., 1998). It was clearly shown in Figure 2 that the lowest IC50 value was obtained from propolis extracts which were collected in the months of the inter-monsoon season, where the lowest is in July and August. From this observation, we anticipate that July and August are the most productive months throughout the year for harvesting propolis extracts from stingless bees that are domesticated in rubber smallholdings surroundings area. Propolis extracts exhibited various chemical compounds and its antioxidant activities, which originated from the different geographical origins of H.itama-FA and H.itama-HB. This observation suggested a need to plant many other species of wild trees, climbing plants and undergrowth, to provide good assimilation of the surroundings with rubber trees, as shown in Table 2.

CONCLUSION

Propolis extract exhibited by different botanical origins of H.itama-FA and H.itama-BH affects the chemical composition of TPC, TFC, and antioxidant activity. Overall, our preliminary findings could portray some benefits and value-added insight into domesticated stingless bees in rubber smallholding and useful guidelines for obtaining propolis extracts during suitable harvesting seasons.

ACKNOWLEDGMENT

I would like to acknowledge the Faculty of Ocean Engineering Technology and Informatics, Central Laboratory, University Malaysia Terengganu, for the equipment used in this study. This work is fully supported by Fundamental Research Grant Scheme (FRGS) FRGS/1/2016/TK02/UMT/01/1, Ministry of Higher Education (MOHE) by the approved fund.

REFERENCES

Abd Jamil, Z., & Zhari, I. (2016). Antibacterial and phenolic content of propolis produced by two Malaysian stingless bees, Heterotrigona itama and Geniotrigona thoracica. *International Journal of Pharmacognosy and Phytochemical Research*, *8*(1), 156–161.

Abd Razak, S. B., Aziz, A. A., Ali, N. A., Ali, M. F., & Visser, F. (2016). The sustainable integration of meliponiculture as an additional income stream for rubber smallholders in Malaysia. *Proceeding in CRI & IRRDB International Rubber Conference*, 21–22.

Al-Hatamleh, M. A. I., Boer, J. C., Wilson, K. L., Plebanski, M., Mohamud, R., & Mustafa, M. Z. (2020). Antioxidant-based medicinal properties of stingless bee products: Recent progress and future directions. *Biomolecules*, *10*(6), 1–28. doi:10.3390/biom10060923 PMID:32570769

Ali, M. F., Abdul Aziz, A., & Williams, A. (2020). Assessing yield and yield stability of hevea clones in the southern and central regions of Malaysia. *Agronomy (Basel)*, *10*(5), 1–15. doi:10.3390/agronomy10050643

Awang, N., Ali, N., Abd Majid, F. A., Hamzah, S., & Abd Razak, S. B. (2018). Total flavonoids and phenolic contents of sticky and hard propolis from 10 species of Indo-Malayan stingless bees. *The Malaysian Journal of Analytical Sciences*, *22*(5), 877–884.

Biesmeijer, J. C., & Ermers, M. C. W. (1999). Social foraging in stingless bees: How colonies of Melipona fasciata choose among nectar sources. *Behavioral Ecology and Sociobiology*, *46*(2), 129–140. doi:10.1007002650050602

Blois, M. S. (1958). Antioxidant determinations by the use of a stable free radical. *Nature*, *181*(4617), 1199–1200. doi:10.1038/1811199a0

Castaldo, S., & Capasso, F. (2002). Propolis, an old remedy used in modern medicine. *Fitoterapia*, *73*, S1–S6. doi:10.1016/S0367-326X(02)00185-5 PMID:12495704

Castro, M. L., Cury, J. A., Rosalen, P. L., Alencar, S. M., Ikegaki, M., Duarte, S., & Koo, H. (2007). Própolis do sudeste e nordeste do Brasil: Influência da sazonalidade na atividade antibacteriana e composição fenólica. *Quimica Nova*, *30*(7), 1512–1516. doi:10.1590/S0100-40422007000700003

Cicco, N., & Lattanzio, V. (2011). The Influence of Initial Carbonate Concentration on the Folin-Ciocalteu Micro-Method for the Determination of Phenolics with Low Concentration in the Presence of Me-thanol: A Comparative Study of Real-Time Monitored Reactions. *American Journal of Analytical Chemistry*, *2*(7), 840–848. doi:10.4236/ajac.2011.27096

Cunha, I., Sawaya, A. C. H. F., Caetano, F. M., Shimizu, M. T., Marcucci, M. C., Drezza, F. T., Povia, G. S., & Carvalho, P. de O. (2004). Factors that influence the yield and composition of Brazilian propolis extracts. *Journal of the Brazilian Chemical Society*, *15*(6), 964–970. doi:10.1590/S0103-50532004000600026

do Nascimento, D. L., & Nascimento, F. S. (2012). Extreme effects of season on the foraging activities and colony productivity of a stingless bee (Melipona asilvai Moure, 1971) in Northeast Brazil. *Psyche*, 2012.

Fadzli, M. H., Jaafar, T. N. A. M., Ali, M. S., Nur, N. F. M., Tan, M. P., & Piah, R. M. (2022). Reproductive aspects of the coastal trevally, Carangoides coeruleopinnatus in Terengganu Waters, Malaysia. *Aquaculture and Fisheries*, *7*(5), 500–506. doi:10.1016/j.aaf.2022.04.009

Jo, W. S., Yang, K. M., Park, H. S., Kim, G. Y., Nam, B. H., Jeong, M. H., & Choi, Y. J. (2012). Effect of microalgal extracts of Tetraselmis suecica against UVB-induced photoaging in human skin fibroblasts. *Toxicological Research*, *28*(4), 241–248. doi:10.5487/TR.2012.28.4.241 PMID:24278616

Kumazawa, S., Hamasaka, T., & Nakayama, T. (2004). Antioxidant activity of propolis of various geographic origins. *Food Chemistry*, *84*(3), 329–339. doi:10.1016/S0308-8146(03)00216-4

Marcucci, M. C. (1995). Propolis: Chemical composition, biological properties and therapeutic activity. *Apidologie*, *26*(2), 83–99. doi:10.1051/apido:19950202

Medic-Saric, M., Bojic, M., Rastija, V., & Cvek, J. (2013). Polyphenolic profiling of Croatian propolis and wine. *Food Technology and Biotechnology*, *51*(2), 159–170.

Mercan, N. (2006). Antimicrobial activity and chemical compositions of Turkish propolis from different regions. *African Journal of Biotechnology*, *5*(11), 1151–1153.

Mokhtar, S. U. (2019). Comparison of total phenolic and flavonoids contents in Malaysian propolis extract with two different extraction solvents. *International Journal of Engineering Technology and Sciences*, *6*(2), 1–11. doi:10.15282/ijets.v6i2.2577

Oldroyd, B. P., & Wongsiri, S. (2009). *Asian honey bees: biology, conservation, and human interactions*. Harvard University Press. doi:10.2307/j.ctv2drhcfb

Oliveira, R. N., Mancini, M. C., Oliveira, F. C. S., de, Passos, T. M., Quilty, B., & Thiré, R. M. (2016). FTIR analysis and quantification of phenols and flavonoids of five commercially available plants extracts used in wound healing. *Matéria (Rio de Janeiro)*, *21*(3), 767–779. doi:10.1590/S1517-707620160003.0072

Panche, A. N., Diwan, A. D., & Chandra, S. R. (2016). Flavonoids: An overview. *Journal of Nutritional Science*, *5*(e47), 1–15. doi:10.1017/jns.2016.41 PMID:28620474

Ponnuchamy, R., Bonhomme, V., Prasad, S., Das, L., Patel, P., Gaucherel, C., Pragasam, A., & Anupama, K. (2014). Honey pollen: Using melissopalynology to understand foraging preferences of bees in tropical South India. *PLoS One*, *9*(7), 1–11. doi:10.1371/journal.pone.0101618 PMID:25004103

Priyadarshan, P. M. (2011). *Biology of Hevea rubber*. Springer. doi:10.1079/9781845936662.0000

Rao, P. S., & Vijayakumar, K. R. (1992). Climatic requirements. In *Developments in crop science* (Vol. 23, pp. 200–219). Elsevier.

Sánchez-Moreno, C., Larrauri, J. A., & Saura-Calixto, F. (1998). A procedure to measure the antiradical efficiency of polyphenols. *Journal of the Science of Food and Agriculture*, *76*(2), 270–276. doi:10.1002/(SICI)1097-0010(199802)76:2<270::AID-JSFA945>3.0.CO;2-9

Slaa, E. J., Chaves, L. A. S., Malagodi-Braga, K. S., & Hofstede, F. E. (2006). Stingless bees in applied pollination: Practice and perspectives. *Apidologie*, *37*(2), 293–315. doi:10.1051/apido:2006022

Suhaizan, L., Norezienda, A., Shamsul, B., Nurul, F. I., & Iffah, H. (2017). Composition and identification of pollen collected by stingless bee (Heterotrigona itama) in forested and coastal area of Terengganu, Malaysia. *Malaysian Applied Biology Journal*, *46*(3), 227–232.

Sujan, M. A., Atim, A. B., & Yaakob, A. M. (1984). Potensi penghasilan madu lebah di kawasan tanaman getah sebagai hasil samping. Siaran Pekebun.

Tajuddin, I. (1986). Integration of animals in rubber plantations. *Agroforestry Systems*, *4*(1), 55–66. doi:10.1007/BF01834702

Zaki, N. N. M., & Razak, A. S. B. (2018). Pollen profile by stingless bee (Heterotrigona itama) reared in rubber smallholding environment at Tepoh, Terengganu. *Malaysian Journal of Microscopy*, *14*(1), 38–54.

Chapter 12
Propagation of Stingless Bees Using a Colony Split Technique for Sustainable Meliponiculture

Shamsul Bahri Abd Razak
Universiti Malaysia Terengganu, Malaysia

Muhammad Izzhan
Universiti Malaysia Terengganu, Malaysia

Nur Aida Hashim
Universiti Malaysia Terengganu, Malaysia

Norasmah Basari
Universiti Malaysia Terengganu, Malaysia

ABSTRACT

Stingless bee farming or meliponiculture is a flourishing industry in Malaysia. The common practice by local stingless bee keepers in order to get new colonies is to obtain feral stingless bees hive from their natural habitat. This practice includes cutting down whole trees to extract stingless bee colonies for domestication. This is not a sustainable way of meliponiculture. The more efficient, sustainable, economic, and eco-friendly method is to breed stingless bees in hives and propagate them for colony multiplication. The aim of this experiment is to provide a good propagation method for stingless bee (Heterotrigonaitama) by dividing brood and queen cells and transfer them into a new box (split method). This method requires a portion of brood with queen cells from original log hive to be transferredinto an empty box hive. From this experiment, 80% of new box hives become new colonies (with new queens).

DOI: 10.4018/978-1-6684-6265-2.ch012

INTRODUCTION

The Stingless bee is a large group of bees from the Apidae family and order Hymenoptera (Michener & Michener, 1974). Approximately 700 species of stingless bees were recorded, most found in tropical countries (Heard, 1999). Each species has unique characteristics regarding morphology and behaviour, including size, population and habitat quality (Fonseca, 2012). Stingless bees are highly eusocial insects. They possessed stingers but were highly reduced, which rendered them unusable for defence. Therefore, they are safe for domestication as they do not sting. In addition, most species are perennial, which suited well for meliponiculture.

Heterotrigonaitama is the common species for its honey and other bee products, such as bee bread and propolis. Malaysian beekeepers prefer this species over others since it is less vulnerable to seasonal changes and capable of surviving in harsh environments (Kelly et al., 2014). Heterotrigonaitama makes up 83.2% of the colonies reared in Malaysia. Heterotrigonaitama prefers mild light intensity, climatic conditions, and the vicinity of abundant flora. This species can be easily distinguished from other commonly encountered stingless bees by size and colouration. The colony of H.itama) is usually nested in a trunk of trees and other cavities. Stingless beekeepers normally cut down the trees to get the colony for domestication. This practice of cutting down trees from the wild could lead to environmental destruction, such as soil erosion, flooding, and global warming.

As in other eusocial insects, the queen of H.itama controls the day-to-day organisation and activities of the colony. The queen is important in transmitting information, and workers' participation in the queen's court was correlated to their activity in cell construction (Sommeijer & De Bruijn 1984). The queen signals her presence to the workers via pheromone. The queen's pheromone comprises mostly volatile compounds originating from the mandibular glands. After fertilisation, the queen can easily be identified through size, especially the engorged abdomen. The queen of stingless bees lays fertile eggs that hatch and become female workers, while infertile eggs will turn into drones or males. The larva that consumed more food will turn into queens compared to the ones that eat less (which will turn into female/male caste) (Hartfelder & Engels, 1989). Stingless bees usually produce new queens only when a daughter colony is to be split off from the mother colony and in the case of accidental loss or replacement of an old queen (Michener & Michener, 1974).

The drone of a stingless bee can be seen from a distinctive congregation of up to several hundred individuals, which can persist several times. Generally, male production in a social insect colony is influenced by outside factors related to climatic periodicity and factors inside the colony, such as colony strength and demographic composition. The stingless bee worker is a female bee with a non-functional ovary. In some bee species, workers can produce drones, but this adaptation may create a conflict between the queens. Workers build cells and provision them with larval food, the queen oviposits, and workers seal the cells immediately (Sakakura et al., 1982). The worker bees collect resin from different plants for construction and defend their nest (Armbruster, 1984). Resin is the material that is sticky, usually white or brown, that worker bees collect from specific plants. Worker bees also collect resin to build propolis. Nogueira-Neto (1997) stated that the worker bee takes a small amount of resin and sticks them on the intruder's body until the legs, antennae and wing are entangled.

The nest structures of stingless bees are more complex than those of Apismelifera. The nest of stingless bees is built within protective cavities, cracks, and crevices, such as in trees and buildings. While the sizes of log hives vary, depending on the cavities of tree trunks. Several types of hives for stingless bees have been described. The Nogueiro-Netohive was mentioned in a publication 50 years ago. The publication also described stingless bee biology and the type of hives. The hives should allow the honey to be harvested effectively without damage or destruction of pollen pots which are generally constructed by stingless bees intermixed with the honey pots.

According to Nogueira-Neto (1997), this criterion leads to the specialised compartmentation of a box hive with a brood chamber and honey compartment for easy honey extraction. The box was prepared with wooden strips, which were joined with nails. In general, the propagation of stingless bees could be categorised into three groups which are the splitting method, swarming technique, and induction technique. This experiment will focus on the splitting method, which involves transferring part of the brood with queen cells of Heterotrigonaitama into an empty hive. The success rate of this experiment was determined by parameters such as the weight of the new colony, the growth of new brood cells, the time of brood cells hatching into a new queen, and the length of the entrance tube.

METHODOLOGY

Sampling Site

This study was conducted at stingless be a farm, Universiti Malaysia Terengganu (UMT) campus (N 5°12' 20.16" E 103° 12' 21.24"), Bukit Kor, Marang, Terengganu from May 2018 to November 2018. The area is suitable for meliponiculture as the plot is surrounded by secondary forest, fruit orchards and agriculture farms that provide abundant food resources for the stingless bees.

Box Hive Design and Dimension

Wooden hives (Figure 1) were constructed to be used as the new hives for new H. itama colonies in this experiment. The hives were made from untreated wooden boards (Hopeaodorata) of 1-inch thickness. The hive consisted of two separate chambers;

1. A wooden box (8-inch x 8inch x 7.5-inch) served as a brood chamber. A hole (0.5' in diameter) that served as the hive entrance was made at the centre of the brood chamber wall and;
2. A compartment (14' x 14' x 5') served as a honey cassette placed above the brood chamber. A wooden cover was made for the honey cassette to prevent pests such as lizards, birds, ants, and beetle from entering the honey cassette. A hole with the size of 0.5' in diameter was drilled in the middle of the honey cassette as the entrance from the brood chamber. A total of 10 wooden hives were set up as replicates.

Figure 1. Box Construction.

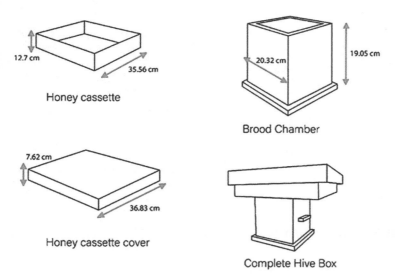

Colony Splitting Method

Ten healthy and thriving H. itama hives containing eggs/brood and queen cells were selected as original (parent) hives. The queen cells were a critical part of the splitting method as this will ensure the survival of the new colony (Ratnieks et al. 2006). Seven layers of brood cells (containing three queen cells) were in the brood chamber in the new hive from each parent hive (Figure 2). The entrance tube (propolis) from the parent hive was cut off and pasted onto the entrance hole of the new hive. It is to attract the workers from the parent hive into the new hive. Some propolis and cerumen materials from the parent hive were also added to the new hive to assist the new colony in constructing its hive. The position of the new hive was then a swap with the parent hive to lure the workers into the new hive further. The same procedures would be repeated for all other nine hives.

Figure 2. Steps of splitting technique

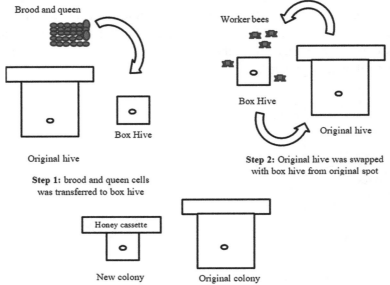

Parameters to Determine the Success of the Split Method

The Weight of the New Colony

Stingless bees require nectar and pollen as their food storage and their growth. Over time, food storage will increase in the new colony/ hive. The weight of the colony in a hive was measured by recording the initial weight of the empty box hive and the final weight of the box hive after seven months. The following formula calculated the increased weight of the new colony:

$$\%WL = \frac{Wf - Wi}{Wi} \times 100 \qquad (1)$$

Where WL is the percentage weight of the colony of the samples, Wf is the final weight of the hive-containing colony, and Wi is the initial weight of the hive without a colony.

Hatching Time of a New Queen

The queen will always move around to check the brood cells, whether the food source is full enough or if any trophic eggs are deposited by workers (Koedam et al., 2007). The time of hatching for a new queen depends on the maturity stage of the queen cells supplied to the new colony. One parameter to measure the new colony's success is when the queen successfully emerges and survives. In this study, daily observation of the queen cells was made to determine the time for the queen's emergence. The monthly observation was also made to determine the survival of the new queen.

The Growth of Brood Cells in the New Colony

Brood refers to the eggs, larvae, and pupae of stingless bees. The size and number of brood cell layers indicate the colony's strength. The growth of the new colony was measured by the length of the new brood cell layers developed after the splitting process. The growth of the brood cell was classified under three different categories: high 6.5-7.5 cm, medium 4.0-6.0 cm, and low 0.3-5 cm.

The Length of the Entrance Tube

The length of the hive entrance tube for each new hive was measured once a month for seven months. Nest entrance architecture is species-specific and highly diverse in stingless bees (Camargo, 1970). According to Biesmeijer et al. (2005), the nest entrances are related to defence, foraging and physio-chemical regulation. However, the structure of the nest entrance or thickness of the resin enclosing the internal nest indicated the age of the nest, bee genetics, and microenvironment, including predators, parasites, rain, and sun (Roubik, 2006).

Statistical Analyses

The Chi-square test for goodness of fit was used to analyse the difference in colonies' weight, brood growth, and tube entrance length. The test was performed in a statistical package, IBM SPSS version 20.

RESULTS

The Weight of the New Colony

Figure 3 shows the weight increases of the new colony (heavy > 2 kg, medium < 2-1.8 kg, and light 1.5 kg) from May until November for 10 colonies of Heterotrigonaitama. Only three hives (30%) recorded heavyweight colony category (ranging from 2.1kg to 2.8 kg), while others fell under medium and light growth weight categories. The percentage number of colonies recorded heavyweight was significantly different compared to medium and lightweight (x=22.460, df=2, p<0.05).

Figure 3. Growth performance of H. itama new colonies by increased weight categories (N=10)

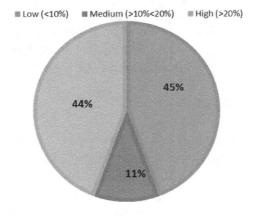

The Growth of Brood Cells in the New Colony

Figure 4 shows the significant growth of brood cells (high 6.5-7.5 cm, medium 4.0-6.0 cm, and low 0.3-5 cm) from May until November. The result showed that three out of the ten new colonies exhibited the highest brood cell growth, which was about 6.5-7.5 cm in height—each after 3 months. Figure 5 shows the colony with the highest growth filled rigorously in the brood chamber with hone and pollen pots. The number of colonies recorded high, medium, and low brood growth was not/significantly different (x=14.000, df=2, p<0.05).

The Length of the Entrance Tube

Figure 7 shows that the length of the entrance tube for the 10 new colonies varied after 7 months. Three hives had been built over 2 cm in length of the entrance tube. Another three colonies had tube lengths from 1 to 2 cm, while the remaining four hives built only short entrance tube lengths of less than 1 cm. The number of colonies recorded long, medium and short tube lengths were not significantly different (x=2.000, df=2, p>0.05).

Figure 4. The percentage growth of the new brood cells

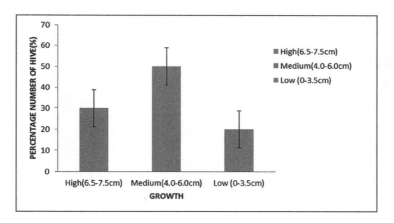

Figure 5. The growth stages of the new colony

Figure 6. Hatching time of queen cell

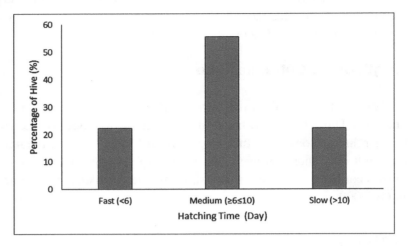

Figure 7. The percentage of hive based on the entrance tube length

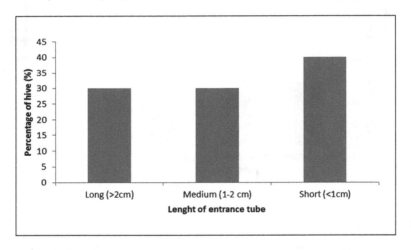

DISCUSSION

The Weight of the New Colony

The productivity of stingless bee colonies is highly dependent on the queen bees' quality. The weight of a newly emerged queen is considered a reliable criterion to evaluate the quality of a stingless bee queen Zeedan (2002) and Taha & Alqarni (2013). Figure 3 shows that 44% of the new colonies filled their colonies vigorously with propolis. The strength of the colonies was influenced by several factors, such as the brood cells stage when doing a splitting technique and the colony management after the splitting process. A colony has a single queen, which is the only productive female. In addition, a healthy colony also has brood, the collective term for eggs, larvae, and pupae. The ability of the colony to succeed depends on several factors, including food availability such as nectar, pollen, water, and the factor related to the geographic distribution of H.itama species (Batista et al., 2003). The weight can be determined by weighing the colony's whole content, consisting of pot honey, brood cell, propolis, bee bread and stingless bees.

The development from brood to adult stingless bees will increase the number of stingless bees in the colony and thus will directly increase the weight of the colony. The quality of the queen also plays an important role in the ability to lay eggs in the colony. More eggs will produce more stingless bees. A previous study by da Veiga (2012) discovered that a colony fed with artificial food resulted in a larger stingless bee; this will also contribute to the size of the whole colony. Moreover, the queens reared from eggs will become larger and heavier (Woyke, 1971), contributing to the colony's positive growth over time. During a normal year, a colony's lowest population of adult workers is in January and February, which is the unproductive season for the bees. Still, the colony can grow to a maximum number of workers in the foraging season (March to October). The time of the year can also be a determining factor in the size of each colony. Dodologlu et al. (2004) reported that the weight of freshly emerged gynes in queen-right colonies was heavier than in queen-less colonies. Still, according to Prinfolosirea (1965), queens that emerged in queen-less colonies were slightly heavier than those reared in queen-right colonies.

In strong colonies, the proportion of stingless bees old enough to forage increases, but smaller and weaker colonies send out a smaller percentage of their bees as foragers (Sagili & Burgett, 2011). Small colonies can be expanded and increased rapidly in size, especially after the rainy season is over and the flowers start to bloom. The over-expanded population of the colony then swarmed to occupy the new nesting sites (Chinhinst et al., 2003). This natural phenomenon in stingless bee propagation will also affect the colony's strength (and the colony's weight). The greater value of heavy colonies can also be illustrated by the amount of honey and bee bread produced by stingless bees. A colony's ability to store surplus honey directly results from the number of stingless bees foraging and the amount of forage in each colony.

The Growth of Brood Cells in the New Colony

The size of brood cells contributed to the vigorous activity of the colony in the new hive (see Figure 4). The brood cell size increases the colony's population by its sheer number. It was also observed that a height of 4.0-6.0 cm gave a more vigorous activity and development to the new colony. The oval brood cells were arranged in multilayers and spiral orientation. The new layers of brood cells were arranged upwards. The queen will lay eggs in empty brood cells prepared by the workers. Nests differed from the single brood cells constructed by the stingless bees to the massive nests containing tens of thousands of specialised cells. The larger drone is less conspicuous than the worker brood cells. According to Lusby (1996), smaller cells positively affect the heat regulation in the brood nests and increases the rate of development of worker bees and their number in the brood nest, favouring their hygienic behaviours. The queen is an important factor in determining whether workers will be reproductively active in colonies or whether the workers will be allowed to build drone cells and lay fertile eggs in those specialised brood cells (Ratnieks et al., 2006). The diameter and length of the brood cell must be large enough to allow the formation of the provision mass as well as the movements of the larvae.

Time of Hatching of a New Queen

The survival of colonies of stingless bees depends on strategies to produce virgin queens (gynes) to ensure an egg-laying queen, whether in a new colony. Figure 6 shows that the time for a new queen to hatch differs between hives because the brood cell stages were not uniform during the transfer of queen brood cells from the log hives. It is quite difficult to determine the stage of brood cells. Different species of stingless bees were reported to have different queen production strategies. This strategy was influenced by the ecological environment and species dependant (Klaus Hartfelder et al., 2006). One of the criteria for determining the success of a newly developed colony was the formation of new brood cells. Fertilised queens will immediately lay eggs in the new colony; these nicely arranged brood cells are a good indicator for the new colony to survive.

In stingless bees, two or more queens can lay eggs in the same nest (Ribeiro et al., 2006). So far, there is no report of such observation in Malaysian Stingless bee species. New queens are produced regularly, but most of them are killed and never allowed to produce eggs. In the case of Heterotrigonaitama, the queen cells are a little larger and situated in the middle of the brood cells. Various studies have shown a large range in the reproductive potential of stingless bee queens. It is measured by their size, ovariole number and spermatheca diameter (Tarpy et al., 2011). Some queens may live for 3-7 years. Stingless bee queens can provide 10-100 cells with eggs daily, depending on the species. Stingless bees produce genes throughout the year (Blochtein & Serrão, 2013), with larvae receiving mass production in the brood cell from its development. These factors support the absence of production of emergency queens in Meliponini, which decreases the chances of colony survival when no genes are available for replacement.

The Length of the Entrance Tube

Nest entrances of stingless bees vary from species to species, ranging from simple holes to dome or trumpet-shaped entrances. However, the unique variation and architecture of entrances for each stingless bee species could not be used as a sole criterion for stingless bee taxonomy. However, they could be a good indicator of the colony's well-being and strength. Couvillon et al. (2008) stated that a strong correlation between relative entrance area and traffic between the hive strongly suggests a trade-off between traffic and security and a significant trend for higher forager traffic. The size of the stingless bee's entrance tube indicated the number of resources needed to build the tube (Roulston & Goodell, 2011). The nest entrances and architecture are closely related to the defence ability and foraging behaviour of stingless bees (Biesmeijer et al., 2005).

There are several ways the nest can be defended which were swarming over the intruder, biting using mandibles, ejection of a burning liquid from the mandibles' camouflage of nesting sites, restricted nest entrances, walls of sticky resin around the entrance tube, and positioning of guard bees in front of nest entrance (Wilson, 1971) and (Wittmann, 1985). The structures of the nest entrance thickness or length of the resin enclosing the internal nest depend on the microenvironment (Roubik, 2006). The availability of food and nesting sites may limit the density of stingless bees in building the nest entrance (Eltz et al., 2002). (Kerr et al., 2010) pointed out that Meliponina bees are threatened by the following factors: habitat destruction by deforestation and fires, small reserves, which may be insufficient to assure genetically viable populations destruction of colonies for honey extraction.

CONCLUSION

In conclusion, this study attempted to propagate stingless bees (H.itama) by dividing brood cells. The experiments showed new box hives exhibited positive development via observing the colony's weight, brood cell growth, the presence of a new queen and the length of the entrance tube as indicators for propagation success. However, These indicators are not limited to these parameters and could be elaborated more. This experiment showed that using a split technique could produce new stingless bee colonies, thus ensuring the sustainability of meliponiculture.

ACKNOWLEDGMENT

We would like to thank the reviewers for all their constructive comments and suggestions.

REFERENCES

Armbruster, W. S. (1984). The role of resin in angiosperm pollination: Ecological and chemical considerations. *American Journal of Botany, 71*(8), 1149–1160. doi:10.1002/j.1537-2197.1984.tb11968.x

Batista, M. A., Ramalho, M., & Soares, A. E. E. (2003). Nesting sites and abundance of Meliponini (Hymenoptera: Apidae) in heterogeneous habitats of the Atlantic Rain Forest, Bahia, Brazil. *Lundiana: International Journal of Biodiversity, 4*(1), 19–23.

Biesmeijer, J. C., Giurfa, M., Koedam, D., Potts, S. G., Joel, D. M., & Dafni, A. (2005). Convergent evolution: Floral guides, stingless bee nest entrances, and insectivorous pitchers. *Naturwissenschaften, 92*(9), 444–450. doi:10.100700114-005-0017-6 PMID:16133103

Blochtein, B., & Serrão, J. E. (2013). Seasonal production and spatial distribution of Melipona bicolor schencki (Apidae; Meliponini) castes in brood combs in southern Brazil. *Apidologie, 44*(2), 176–187. doi:10.100713592-012-0169-2

Camargo, J. M. F. (1970). Nihos e biologia de algumas especies de Meliponideos (Hymenoptera: Apidae) de regiao de Porto Velho, Territorio de Rondonia, Brasil. *Revista de Biología Tropical, 16*(2), 207–239.

Chinhinst, T. X., Grob, G. B. J., Meeuwsen, F. J. A. J., & Sommeijer, M. J. (2003). Patterns of male production in the stingless bee Melipona favosa (Apidae, Meliponini). *Apidologie, 34*(2), 161–170. doi:10.1051/apido:2003008

Couvillon, M. J., Wenseleers, T., Imperatriz-Fonseca, V. L., Nogueira-Neto, P., & Ratnieks, F. L. W. (2008). Comparative study in stingless bees (Meliponini) demonstrates that nest entrance size predicts traffic and defensivity. *Journal of Evolutionary Biology, 21*(1), 194–201. doi:10.1111/j.1420-9101.2007.01457.x PMID:18021200

da Veiga, J. E. (2012). *O desenvolvimento agrícola: uma visão histórica.* Edusp.

Dodologlu, A., Emsen, B., & Gene, F. (2004). Comparison of some characteristics of queen honey bees (Apis mellifera L.) reared by using Doolittle method and natural queen cells. *Journal of Applied Animal Research, 26*(2), 113–115. doi:10.1080/09 712119.2004.9706518

Eltz, T., Brühl, C. A., Van der Kaars, S., & Linsenmair, E. K. (2002). Determinants of stingless bee nest density in lowland dipterocarp forests of Sabah, Malaysia. *Oecologia, 131*(1), 27–34. doi:10.100700442-001-0848-6 PMID:28547507

Fonseca, V. L. I. (2012). *Best management practices in agriculture for sustainable use and conservation of pollinators.* Academic Press.

Hartfelder, K., & Engels, W. (1989). The composition of larval food in stingless bees: Evaluating nutritional balance by chemosystematic methods. *Insectes Sociaux, 36*(1), 1–14. doi:10.1007/BF02225876

Hartfelder, K., Makert, G. R., Judice, C. C., Pereira, G. A. G., Santana, W. C., Dallacqua, R., & Bitondi, M. M. G. (2006). Physiological and genetic mechanisms underlying caste development, reproduction and division of labor in stingless bees. *Apidologie, 37*(2), 144–163. doi:10.1051/apido:2006013

Heard, T. A. (1999). The role of stingless bees in crop pollination. *Annual Review of Entomology, 44*(1), 183–206. doi:10.1146/annurev.ento.44.1.183 PMID:15012371

Kelly, N., Farisya, M. S. N., Kumara, T. K., & Marcela, P. (2014). Species Diversity and External Nest Characteristics of Stingless Bees in Meliponiculture. *Pertanika. Journal of Tropical Agricultural Science, 37*(3), 293–298.

Kerr, W. E., Carvalho, G. A., Silva, A. C., da, & Assis, M. da G. P. de. (2010). Aspectos pouco mencionados da biodiversidade amazônica. *Parcerias Estratégicas, 6*(12), 20–41.

Koedam, D., Aponte, O. I. C., & Imperatriz-Fonseca, V. L. (2007). Egg laying and oophagy by reproductive workers in the polygynous stingless bee Melipona bicolor (Hymenoptera, Meliponini). *Apidologie, 38*(1), 55–66. doi:10.1051/apido:2006053

Lusby, D. A. (1996). Small cell size foundation for mite control. *American Bee Journal, 136*(7), 468–470.

Michener, C. D., & Michener, C. D. (1974). *The social behavior of the bees: a comparative study.* Harvard University Press.

Nogueira-Neto, P. (1997). *Vida e criação de abelhas indígenas sem ferrão.* Academic Press.

Prinfolosirea, M. E. C. M. O. (1965). Diferitelor Metode de Pregatire a Naterialului biologic. *Lucr Stiint Stat Cont Seri Apic, 6*, 15–21.

Ratnieks, F. L. W., Foster, K. R., & Wenseleers, T. (2006). Conflict resolution in insect societies. *Annual Review of Entomology, 51*(1), 581–608. doi:10.1146/annurev. ento.51.110104.151003 PMID:16332224

Ribeiro, M. de F., Wenseleers, T., Santos Filho, P. S., & Alves, D. A. (2006). Miniature queens in stingless bees: Basic facts and evolutionary hypotheses. *Apidologie, 37*(2), 191–206. doi:10.1051/apido:2006023

Roubik, D. W. (2006). Stingless bee nesting biology. *Apidologie, 37*(2), 124–143. doi:10.1051/apido:2006026

Roulston, T. H., & Goodell, K. (2011). The role of resources and risks in regulating wild bee populations. *Annual Review of Entomology, 56*(1), 293–312. doi:10.1146/annurev-ento-120709-144802 PMID:20822447

Sagili, R. R., & Burgett, D. M. (2011). *Evaluating honey bee colonies for pollination: a guide for commercial growers and beekeepers.* Academic Press.

Sakakura, T., Sakagami, Y., & Nishizuka, Y. (1982). Dual origin of mesenchymal tissues participating in mouse mammary gland embryogenesis. *Developmental Biology, 91*(1), 202–207. doi:10.1016/0012-1606(82)90024-0 PMID:7095258

Sommeijer, M. J., & De Bruijn, L. L. M. (1984). Social Behaviour of Stingless Bees:" Bee-Dances" by Workers of the Royal Court and the Rhythmicity of Brood Cell Provisioning and Oviposition Behaviour. *Behaviour, 89*(3–4), 299–315. doi:10.1163/156853984X00434

Taha, E. K. A., & Alqarni, A. S. (2013). Morphometric and reproductive organs characters of Apis mellifera jemenitica drones in comparison to Apis mellifera carnica. *International Journal of Scientific and Engineering Research, 4*(10), 411–415.

Tarpy, D. R., Keller, J. J., Caren, J. R., & Delaney, D. A. (2011). Experimentally induced variation in the physical reproductive potential and mating success in honey bee queens. *Insectes Sociaux, 58*(4), 569–574. doi:10.100700040-011-0180-z

Wilson, E. O. (1971). *The insect societies.* Cambridge, Massachusetts, USA, Harvard University Press [Distributed by…. Wittmann, D. (1985). Aerial defense of the nest by workers of the stingless bee Trigona (Tetragonisca) angustula (Latreille) (Hymenoptera: Apidae). *Behavioral Ecology and Sociobiology, 16*(2), 111–114.

Woyke, J. (1971). Correlations between the age at which honeybee brood was grafted, characteristics of the resultant queens, and results of insemination. *Journal of Apicultural Research*, *10*(1), 45–55. doi:10.1080/00218839.1971.11099669

Zeedan, E. W. M. (2002). *Studies on certain factors affecting production and quality of queen honeybees (Apis mellifera L.) in Giza region* [Sc. Thesis]. Fac. Agric. Cairo Univ.

Chapter 13
Stingless Bees and Honey Bees of West Sumatra, Indonesia

Siti Salmah
Universitas Andalas, Indonesia

Henny Herwina
Universitas Andalas, Indonesia

Jasmi Jasmi
College of Health Sciences Indonesia, Indonesia

Idrus Abbas
Universitas Andalas, Indonesia

Dahelmi Dahelmi
Universitas Andalas, Indonesia

Muhammad N. Janra
Universitas Andalas, Indonesia

Buti Yohenda Christy
Universitas Andalas, Indonesia

ABSTRACT

This chapter summarizes the works on Sumatran bees from three research periods: between 1980-1987 on several locations in West Sumatra, 1990 at Kerinci Seblat National Park, and between 2019-2020 at some beekeepers in West Sumatra. In total, there were 27 stingless bee species, one stingless bee forma (Tetragonula minangkabau forma darek), and three honey bee species identified. Most of these stingless bee and honey bee species inhabit the Sumatran lowland primary forest.

DOI: 10.4018/978-1-6684-6265-2.ch013

There were four patterns of species distribution observed in this study: rare species that were confined to primary forest, moderate or abundant species that were bound to primary forest, species that inhabited both primary and secondary forest, and species that adapt to disturbed areas. Apis andreniformis, A. dorsata, A. cerana indica, Heterotrigona itama, Sundatrigona moorei, Tetragonula fuscobaltealta, T. drescheri, T. laeviceps, and T. minangkabau were example of adaptive species.

INTRODUCTION

There are three tribes of Apidae bees in Southeast Asia: the stingless bees of Meliponini tribe, the honey bees from the Apini tribe and the bumblebees of the Bombini tribe (Maa, 1953; Sakagami, 1975, 1978; Schwarz, 1948). Honey bees and stingless bees are not only considered ecologically important, but they also produce honey. In addition, some valuable bee products such as wax, royal jelly and propolis (resin or cerumen) are also yielded. Recently, there has been an increasing interest in using pollen harvested from beehives as part of the human diet. European athletes were among the first to take advantage of using pollen as a food supplement, and the positive effects on performance were reported. Nowadays, pollen is used in two general ways. Europeans, Russians and Americans consider it a "healthy food" with similar efficacy as the general tonic. In Europe and Russia, Pollen is also used to clinically treat chronic prostatitis, bleeding stomach ulcers, respiratory infections, and allergy reactions. Hence, the demand for pollen has remarkably increased, urging the active involvement of beekeepers in collecting and trading pollen (Shimizu & Morse, 2018).

The attention has now shifted to stingless bee products, especially as the honeybee's demand cannot be supplied alone. It is supported by various studies that addressed medical properties contained in stingless bee's products. Its honey contains many important compounds, including tannins, flavonoids, coumarin and carbohydrates (Chuttong et al., 2016). Meanwhile, its bee pollen contains bioactive compounds such as proteins, amino acids, lipids, carbohydrates, minerals, vitamins, and polyphenols that are essential for treating metabolic disorders (Khalifa et al., 2021). The propolis from stingless bee propolis has antibacterial potency against Enterobacter sakazakii (Hasan et al., 2016) and toward Staphylococcus aureus and Eschericia coli (Yusop et al., 2018). In Indonesia, meliponiculture started gaining momentum around 2015, much later than apiculture.

Sumatran stingless and honey bees make up to 74% of the total known pollinator insects for various flowering plants on the island (Inoue & Salmah, 1990); it was similarly observed in the Malay Peninsula (Yoshikawa, 1969). Concerning the emergence of rampant meliponiculture activity in Sumatra, there are challenges in preserving the diversity of stingless and honey bees therein. It is common for people in Sumatra to extract natural resources from the forest and other natural landscapes to gain economic value, such as timbers, rattan, plant resin, honey and wax from the giant honey bees Apis dorsata (Veth, 1881). Some beekeepers still rely on taking wild bee colonies to develop their bee farms. The introduction of non-native bees to breed in Sumatran meliponicultures further disturbs the existence of Sumatran native bees. Furthermore, it is complicated by the reduction of primary forest cover from the shifting into various human-made landscapes in the last two decades (Margono et al., 2014). Therefore, conservation measures must be taken to preserve the existence of Sumatran bees. In this paper, we focus on exploring species richness and distribution of stingless and honey bees in various habitat types and altitudes in West Sumatra.

METHODOLOGY

An array of specimen collection is used in this study derived from three research periods. The first period was between September 1980 to June 1987 at 107 localities within the administrative boundary of West Sumatra Province (Figure 1). The second period lasted from July to Desember 1990 at Kerinci Seblat National Park (Figure 2), while the third was performed from April 2019 until August 2020 at 16 beekeeping sites in West Sumatra (Figure 3). The study used collection techniques such as net sweeping on insects visiting flowers (coded as 'fl'), baiting insects with mixture of diluted honey with 40% sugar concentrate sprayed on leaves ('hn'), collecting from nests ('ns'), collecting insects attracted by sweat ('sw') and collecting insects attracted into wetted ash or cement ('cm') (Inoue & Salmah, 1990). The nesting site, especially on the tree, was identified and recorded. Some nests were collected to assess the number of workers, nest volume, the volume of honey, pollen, wax or resin and others produced. Site altitude was divided based on elevational level, comprising 0 - 200 m, 200 - 500 m, 500 - 1000 m, 1000 - 1500 m, and 1500 - 2000 m. Habitats where the collection was conducted, were assessed and categorized as primary forest, secondary forest, or disturbed areas.

Figure 1 displays stingless bee species recorded at sampling localities in West Sumatra and four adjacent provinces during the first research period. Sampling localities from north southward Pasaman Regency (PM), Limapuluh Kota Regency (LK), Agam Regency (AG), Padang Pariaman Regency (PP), Tanah Datar Regency (TD), Sawahlunto Sijunjung Regency (SW), Solok Regency (SL), Padang City (PD) and Pesisir Selatan Regency (PS). The shaded area indicates primary forest (Inoue & Salmah, 1990).

Figure 1. Stingless bee species recorded at sampling localities in West Sumatra

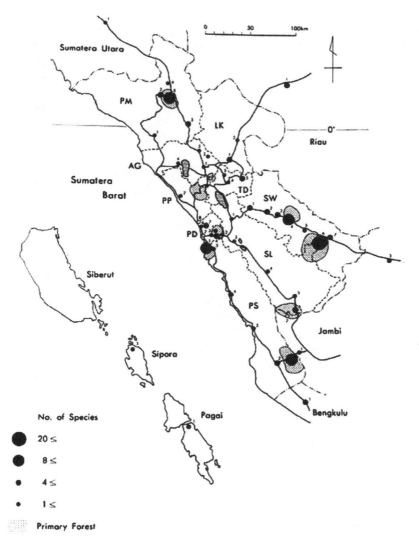

RESULTS AND DISCUSSIONS

This study recorded a total of 27 stingless bee species, one stingless bee forma and three honey bee species from West Sumatra and Kerinci Seblat National Park (Table 1). Chronologically, the species inventory started with 23 stingless bee species, one stingless bee forma and three honey bee species recorded during the first research period. Five stingless bee species were a new record for Sumatra, while Tetragonula

209

Figure 2. Sampling sites in Kerinci Seblat National Park during the second research period. Localities are indicated by 1 = Muaro Sako, Pancung Soal; 2 = Tapan Watas; 3 = Lubuk Gadang, Sangir; 4 = Letter W, Sangir; 5 = Gunung Rasam, Lembah Gumanti; 6 = Gunung Kerinci; 7 = Gunung Tujuh; 8 = Pangkalan, Musi Rawas; 9 = Air Duku, Sambirejo; 10 = Bukit Kelam, Sambirejo; 11 = Penarik, Muko-Muko; 12 = Bukit Setajam, Muko-Muko (SALMAH 1991).

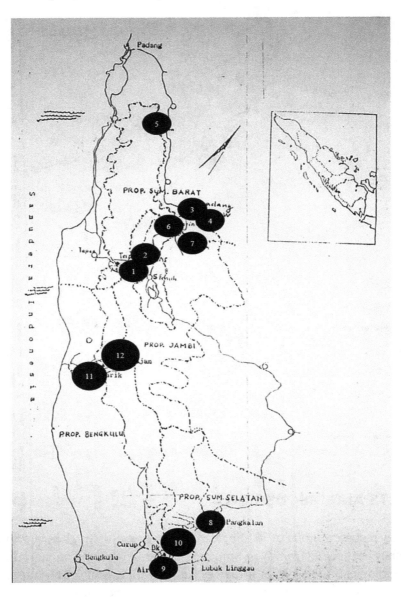

Figure 3. Sampling localities for West Sumatran stingless bees during the third period.
Note: 1 = Biological Education and Research Forest (BERF) of Andalas University, Padang; 2 = Sungai Lasi, Solok; 3 = Kelok Macan, Sawahlunto; 4 = Nagari Lalan, Sijunjung; 5 = Dasawisma Nagari Lalan, Sijunjung; 6 = YF Farm, Sijunjung; 7 = Sungai Dareh, Dharmasraya; 8 = 7 Koto, Padang Pariaman; 9 = Sicincin; 10 = Pariangan, Batusangkar; 11 = Padang Panjang; 12 = Dinas Kehutanan, Payakumbuh; 13 = Sarilamak, Payakumbuh; 14 = Sianok, Bukittinggi; 15 = Solok Selatan; 16 = Ngungun Saok, Lubuk Minturun
Source: Herwina et al. (2021)

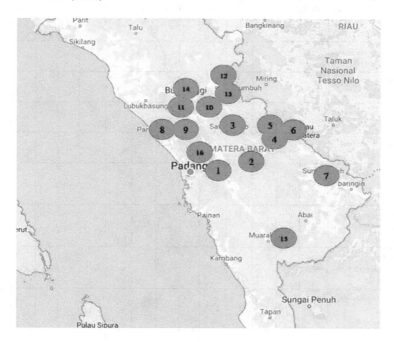

minangkabau and Sundatrigona lieftincki were described as new species. Another stingless bee species, Tetrigona melanoleuca, was recorded during the second research period and added to the species inventory. On the third research period, three more stingless bee species were added as new distribution recorded in Sumatran species inventory, i.e., Tetrigona binghami, Tetragonula geissleri, and Tetragonula testaceitarsis. Out of the 27 stingless bee and honey bee species recorded from West Sumatra, 26 inhibit lowland primary forests.

Only Sundatrigona lieftincki was observed from the primary and secondary forest at 1200 - 1400 m elevation. On the other hand, Pariotrigona pendleburyi and Tetragonula reepeni were found only in the primary forest at 0 - 200 m, while Lisotrigona scintilans were at 0 - 500 m. These 27 bee species can be found in the primary forest, which then reduced to 17 stingless bee species and three honey bee species in the secondary forest. The disturbed areas retained 17 species of both stingless bees and honey bees.

Table 1. Stingless bees and honey bees in West Sumatra, their habitats and altitudinal distribution.

No.	Species	Primary Forest					Secondary Forest					Disturbed Areas				
		0-200 m	200-500	500-1000	1000-1500	1500-2000	0-200 m	200-500	500-1000	1000-1500	1500-2000	0-200 m	200-500	500-1000	1000-1500	1500-2000
1	*Geniotrigona thoracica*	+	+	+	-	-	+	+	-	-	-	+	+	-	-	-
2	*Lepidotrigona nitidiventris*	+	+	-	-	-	+	+	-	-	-	-	-	-	-	-
3	*Lepidotrigona trochanterica**	+	+	+	+	-	-	-	-	-	-	-	-	-	-	-
4	*Lepidotrigona terminata*	+	+	+	-	-	+	-	-	-	-	-	-	-	-	-
5	*Lepidotrigona ventralis*	+	+	+	-	-	-	-	-	-	-	-	-	-	-	-
6	*Lisotrigona scintillans** / *L. cacciae****	+	+	-	-	-	-	-	-	-	-	-	-	-	-	-
7	*Heterotrigona (Heterotrigona) itama*	+	+	+	-	-	+	+	-	-	-	+	+	+	-	-
8	*Heterotrigona (Sundatrigona) moorei**	+	+	-	-	-	+	+	+	+	-	+	-	+	-	-
9	*Heterotrigona (Sundatrigona) lieftincki**	-	-	-	+	-	-	-	-	+	-	-	-	-	-	-
10	*Homotrigona (Homotrigona) fimbriata*	+	+	-	-	-	-	-	-	-	-	+	-	-	-	-
11	*Homotrigona (Lophotrigona) canifrons*	+	+	-	-	-	+	-	+	-	-	+	-	-	-	-
12	*Homotrigona (Tetrigona) apicalis*	+	+	+	-	-	+	+	+	-	-	+	-	+	-	-
13	*Homotrigona (Tetrigona) binghami**	-	-	-	-	-	-	+	-	-	-	-	-	-	-	-
14	*Homotrigona (Tetrigona) melanoleuca**	+	-	-	-	-	+	-	-	-	-	-	-	-	-	-
15	*Pariotrigona pendleburyi**	+	-	-	-	-	-	-	-	-	-	-	-	-	-	-
16	*Tetragonula (Tetragonilla) atripes*	+	+	-	-	-	+	-	-	-	-	+	-	-	-	-

continued on following page

Table 1. Continued

No.	Species	Primary Forest					Secondary Forest					Disturbed Areas				
		0-200 m	200-500	500-1000	1000-1500	1500-2000	0-200 m	200-500	500-1000	1000-1500	1500-2000	0-200 m	200-500	500-1000	1000-1500	1500-2000
17	*Tetragonula (Tetragonilla) collina*	+	+	+	-	-	+	+	-	-	-	+	-	-	-	-
18	*Tetragonula (Tetragonilla) fuscibasis**	+	+	-	-	-	+	+	-	-	-	-	-	-	-	-
19	*Tetragonula (Tetragonula) drescheri*	+	+	+	-	-	+	+	+	-	-	+	+	+	-	-
20	*Tetragonula (Tetragonula) fuscobaltealta*	+	+	-	-	-	+	+	+	-	-	+	+	+	-	-
21	*Tetragonula (Tetragonula) geissleri**	-	-	-	-	-	+	-	-	-	-	-	-	-	-	-
22	*Tetragonula (Tetragonula) laeviceps*	+	+	+	+	-	+	+	+	+	-	+	+	+	-	-
23	*Tetragonula (Tetragonula) l. (small)*	+	+	-	-	-	+	+	+	-	-	+	+	-	-	-
24	*Tetragonula (Tetragonula) melina**	+	+	-	-	-	-	-	-	-	-	-	-	-	-	-
25	*Tetragonula (Tetragonula) minangkabau***	+	+	+	-	-	+	+	-	-	-	+	+	-	-	-
26	*Tetragonula (Tetragonula) m. f.* darek**	+	+	+	-	-	+	+	+	-	-	+	+	+	-	-
27	*Tetragonula (Tetragonula) reepeni**	+	-	-	-	-	-	-	-	-	-	-	-	-	-	-
28	*Tetragonula (Tetragonula) testaceitarsis**	-	-	-	-	-	-	-	+	-	-	-	-	-	-	-
29	*Apis (Micrapis) andreniformis*	-	+	+	-	-	+	-	-	-	-	+	-	-	-	-
30	*Apis (Megapis) dorsata*	+	+	+	-	-	+	+	+	-	-	+	+	-	-	-
31	*Apis (Apis) cerana indica*	+	+	+	+	+	+	+	+	+	-	+	-	-	+	+

Note: * = new record for Sumatra, ** = new species, *** = new name
Source: Grüter (2020)

Primary forests retain higher species richness and relative abundance than the other two habitats (Inoue & Salmah, 1990). Elevation-wise, 25 species of stingless and honey bees were recorded at 0 - 200 m elevation; it drastically decreased to five species at 1000 - 1500 m elevation (Figure 4). All three honey bee species were recorded up until 1000 m elevation, which later only Apis cerana could be found in all elevational ranges (0 - 2500 m elevation). The stingless bee subgenus Lepidotrigona was abundant at the highland of over 800 m elevation. This study revealed four distributional patterns for Sumatran apid bees; rare species seemed to be confined to the primary forest, moderate or abundant species inhibited the primary forest, common species in both primary and secondary forest, and species that adapted to disturbed areas. The latter species included Apis andreniformis, A. dorsata, A. cerana indica, Heterotrigona itama, Tetragonula drescheri, T. fuscobaltealta, T. laeviceps, T. minangkabau and Sundatrigona moorei. After all, the tropic and subtropic areas in the Oriental region have poor apid diversity compared with neotropics. In total, the number of genera and subgenera in this region only reach 89, far behind the 315 taxa found in the Neotropics (Michener, 1979).

Figure 4. Number of stingless bee and honey bee species as a function against elevation and different habitat types

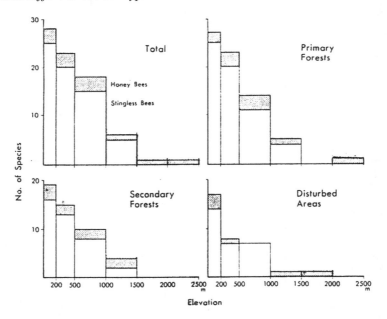

Fourteen units of stingless bee and honey bee nests were collected during the survey in West Sumatra. The species, nest site selection and average nest volumes are shown in Figure 5. Nests that used tree cavities varied from only 0.31 l in T. fuscobaltealta to 3,301 l in Lophotrigona canifrons. Many anthropophilic bee species were observed to build their nests adjacent to human settlements. This anthropophilic stingless bee included T. drescheri, T. fuscobaltealta, T. laeviceps, and T. Minangkabau, while the three honey bee species were also recorded to be anthropophilous. Nesting site selection and tree species used for establishing a nest are shown in Table 2.

Figure 5. Average nest volumes of stingless and honey bees at respective nesting site types.

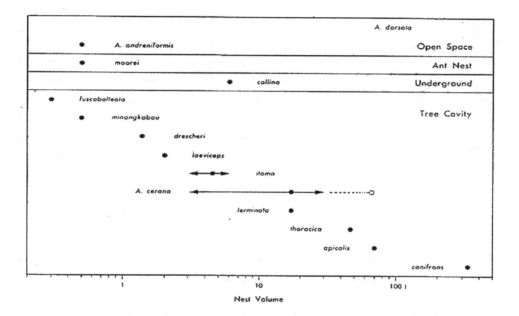

Stingless and honey bees may have specific site requirements for building their nest. Early research on nesting sites for stingless bees in Panama observed exclusivity in nesting behaviour. It was found that 19 stingless bee species used only tree cavities, five solely used termite nests, two nested inside the ground, and four others built exposed nests (Roubik, 1983). On the other hand, seven species nested in tree cavities occasionally used man-made cavities. Nesting sites of A. andreniformis and A. dorsata never overlap with those of stingless bees, probably as these honey bees use spaces under branches at a considerable height of trees. It

Table 2. Nest site selection and tree species used for stingless bee and honey bee nests

Nest Site Selection/ Tree Species	Abbreviation
Bamboo stems	T.f, T.mi, T.l, T.t
Interspaces between wooden walls	T.f, T.mi
Crack or crevices in rock	T.f, H.i
Holes in lime stones	T.f, T.l
Hollow branches of woody fern *Cyathea* spp., *Alsophila* spp.	T.mi, T.l
Tree trunks	T.d
Cracks in house or buildings	T.l, T.d, A.c
Clove *Eugenia aromatica*	T.i, A.c
Coconut *Cocos nucifera*	T.i, T.d, A.c
Buttress of big trees	T.c
Hollow in tree trunks of fig *Ficus* spp.	L.ter, G.th, A.c
Hollow in tree trunks of tarok *Artocarpus elasticus*	T.a, L.can
Crevices in stone walls	T.a
Nest of *Crematogaster* ant built within twigs of shrubs	S.mo
Boulder in the sloping cliff	A.d
In the underside of roof of big building	A.d
In open spaces in storage room	A.c
Undersurface of twigs of shrubs in disturbed areas	A.ad

Note: G.th = G. thoracica, H.i = H. itama, L.ter = L. terminata, L.can = L. canifrons, T.a = T. apicalis, T.f = T. fuscobaltealta, T.mi = T. minangkabau, T.l = T. laeviceps, T.d = T. drescheri, T.c = T. colina, T.t = T. testaceitarsis, S.mo = S. moorei, A.d = A. dorsata, A.c = A. cerana, A.ad = A. andreniformis

is in line with the diversity of apid bees, the diversity of nesting sites in neotropics is much higher than what is observed in Southeast Asia (Sakagami, 1961).

The honey bee nests collected from the studies in West Sumatra were used to count bee workers and to measure the resources procured by each bee colony (Table 3). Resources stored in A. dorsata nests were more than measured in the other two honey bees' nests. The huge number and size of A. dorsata workers significantly increase the volume of honey, bee pollen or wax collected by this honey bee species. As observed elsewhere in Asia, such as India, Thailand or Philippines, the giant honey bees A. dorsata can accumulate up to dozens of kilograms of honey (Ghatge, 1951; Roubik, 1992). With this productivity, A. dorsata became the most harvested honey bee in Sumatra as well, despite the average honey volume stored in Sumatran nests only measuring up to 5 kg (Revilla et al., 1986; Salmah, 1989).

The difference in honey volume between Sumatran and south Asian A. dorsata could be explained by harvesting frequency in either region. Beekeepers in north

216

India and Nepal maintain their harvest only during the monsoon period per year, giving year-round period for the bee colony to collect scarce resources for their nest, resulting in more honey stored (Roubik et al., 1985). This seasonality confinement for harvesting honey is also observed in Sri Lanka (Koeniger & Koeniger, 1980). While in Sumatra, the flowering time of plants and vegetation is rather a continuous process, renders the constant availability of honey in A. dorsata nest and invites more frequent harvesting (Salmah, 1989). Moreover, intense pouring during the rainy season in Sumatra tends to delay honey collection by the bee workers prominently.

Table 3. Measurements on nest resources, number and body length of workers of three honey bee species in West Sumatra

Resource Types	Measurement	: Species		
		: *A. dorsata* (n = 4)	: *A. cerana* (n = 10)	: *A. andreniformis* (n = 10)
Honey	Volume (cm³)	172.2 – 800.4	0.6 – 3,974.0	3.1 – 18.2
	Sugar (%)	73.2 – 76.4	66.6 – 74.0	64.6 – 71.9
	Sugar weight (g)	182.0 – 833.3	0.3 – 1972.0	2.7 – 16.8
Pollen	Total volume (cm³)	135.0 – 1017.1	3.1 – 93.3	0.1 – 6.9
Wax	Total weight (g)	95.0 – 273.1	14.0 – 373.0	3.5 – 7.5
No. of worker		11,613 – 40,004	3,515 – 34,360	3,263 – 5,901
Body length (mm)		14.0 – 19.4	10.8 – 12.5	8.0 – 9.5

In comparison, the same measurements were also conducted on the nest resources and workers of three stingless species in West Sumatra, as shown in Table 4. The resources produced by T. drescheri were measured as superior to what was provided in the nest of T. apicalis and T. minangkabau. The number of honey and pollen pots in a stingless bee nest will depend on the colony condition and age. Meanwhile, the sugar content in stingless bee honey is similar to bee's honey in general. The lesser body of a stingless bee compared to the size of a honey bee consequently impact an apparent less volume of honey it produces. However, the breed of stingless bee in an artificial environment looks promising. Six stingless bee species, e.g., T. minangkabau, T. fuscobaltealta, T. itama, T. moorei, T. laeviceps and T. drescheri, have been successfully reared in the laboratory by using nests equipped with glass-topping apparatus and function to facilitate the observation. This success also means they can be bred more intensively without fully relying on the relocation of wild colonies.

217

Table 4. Measurement of resources, number and body length of workers from three stingless bee species in West Sumatra

Resource Types	Measurements	: Species		
		: *T. drescheri* (n = 4)	: *T. apicalis* (n = 1)	: *T. minangkabau* (n = 10)
Honey	Volume (cm³)	36/54 - 470/705	72	26 – 531
	Sugar (%)	65.0 – 70.0	62.0 – 68.9	71.2 – 74.0
Pollen	Pots (ml)	27/40.5 – 323/484.5	192	0 – 94
Total pots		69 – 795		52 – 555
Volume of nest cavity (cm³)		1,440 – 14,350	70,000	500
Volume of nest (cm³)		180 – 7,490	-	-
No. of worker		261 – 6,232	6,166	503 – 1,692
Length of body (mm)		3.7 – 4.6	6.4 – 7.1	3.2 – 3.5

The second research period lasted from July to December 1990 at Kerinci Seblat National Park and added one new stingless bee species to the inventory. Tetrigona melanoleuca was recorded from the Bengkulu portion of the national park at 120 – 160 m elevation. Identification was made from individuals collected on rambutan (Nephelium lappaceum L.) flowers growing in formerly cleared secondary forests. Stingless bee T. melanoleuca is similar in morphology to T. apicalis, except for its colouration and width of malar space. Stingless bee T. melanoleuca is darker than T. apicalis whilemalar space in the first species is shorter than the width of flagellomere II, with a ratio of 5:7 (Sakagami et al., 1990).

Also, during this period, the stingless bee Geniotrigona thoracica was observed to nest in a residential area known previously as a secondary forest. This stingless bee put its colonies in tree hollow and stone crevices in this disturbed habitat. This stingless bee was previously reported from southern parts of Sumatra (Pedro & Camargo, 2003). Globally, G. thoracica is distributed in Borneo, the Malay Peninsula and Thailand. In the meantime, the stingless bee Heterotrigona itama was found at 1,500 m elevation in secondary forest and disturbed Kerinci Seblat National Park areas. Heterotrigona itama is distributed in Sumatra, Java, Kalimantan, Malaysia, and Thailand (Sakagami et al., 1990). The third research period booked a total of 13 stingless bee species from six genera. The study in this period focused on species bred in meliponicultures and apicultures in West Sumatra. Despite the robustness of results from the previous research periods, this last study recorded three new occurrences of stingless bees; Tetrigona binghami, Tetragonula geissleri and Tetragonula testaceitarsis.

CONCLUSION

The total number of bee species identified from three research periods was 27 stingless bees species, 1 stingless bee forma (Tetragonula minangkabau forma darek), and 3 honey bee species. From first research period (1980-1987), 23 stingless bee species (and 1 forma) and 3 honeybee species were recorded particularly in the lowland primary forest. In second research period (1990), Tetrigona melanoleuca was added to the species inventory. While in the third research period, three more species were recorded, i.e., Tetrigona binghami, Tetragonula geissleri, and Tetragonula testaceitarsis. This study observed four patterns of species distribution: rare species that were confined to the primary forest, moderate or abundant species that were bound to the primary forest, species that inhabited both primary and secondary forests, and species that adapted to disturbed areas.

ACKNOWLEDGMENT

This work was funded by the Fundamental Research Grant Scheme (FRGS/1/2019/ WAB11/UMT/02/2) and Naluri Pantas Sdn. Bhd. The authors would like to extend their gratitude to everyone who helped in the article writing process.

REFERENCES

Chuttong, B., Chanbang, Y., Sringarm, K., & Burgett, M. (2016). Physicochemical profiles of stingless bee (Apidae: Meliponini) honey from South east Asia (Thailand). *Food Chemistry*, *192*, 149–155. doi:10.1016/j.foodchem.2015.06.089 PMID:26304332

Ghatge, A. L. (1951). The bees of India. *Indian Bee J.*, *13*, 88.

Grüter, C. (2020). *Stingless bees*. Springer International Publishing. doi:10.1007/978-3-030-60090-7

Hasan, A. E. Z., Nashrianto, H., Juhaeni, R. N., & Artika, I. M. (2016). Optimization of conditions for flavonoids extraction from mangosteen (Garcinia mangostana L.). *Der Pharmacia Lettre*, *8*(18), 114–120.

Herwina, H., Salmah, S., Janra, M. N., Nurdin, J., & Sari, D. A. (2021). Stingless beekeeping (Hymenoptera: Apidae: Meliponini) and Its Potency for Other Related-Ventures in West Sumatra. *Journal of Physics: Conference Series*, *1940*(1), 12073. doi:10.1088/1742-6596/1940/1/012073

Inoue, T., & Salmah, S. (1990). Nest site selection and reproductive ecology of the Asian honey bee, Apis cerana indica, in central Sumatra. *Nest Site Selection and Reproductive Ecology of the Asian Honey Bee, Apis Cerana Indica, in Central Sumatra.*, 219–232.

Khalifa, S. A. M., Elashal, M. H., Yosri, N., Du, M., Musharraf, S. G., Nahar, L., Sarker, S. D., Guo, Z., Cao, W., Zou, X., Abd El-Wahed, A. A., Xiao, J., Omar, H. A., Hegazy, M.-E. F., & El-Seedi, H. R. (2021). Bee pollen: Current status and therapeutic potential. *Nutrients*, *13*(6), 1876. doi:10.3390/nu13061876 PMID:34072636

Koeniger, N., & Koeniger, G. (1980). Observations and experiments on migration and dance communication of Apsis dorsata in Sri Lanka. *Journal of Apicultural Research*, *19*(1), 21–34. doi:10.1080/00218839.1980.11099994

Maa, T.-C. (1953). *An inquiry into the systematics of the tribus Apidini or honeybees (Hym.)*. Archipel.

Margono, B. A., Potapov, P. V., Turubanova, S., Stolle, F., & Hansen, M. C. (2014). Primary forest cover loss in Indonesia over 2000–2012. *Nature Climate Change*, *4*(8), 730–735. doi:10.1038/nclimate2277

Michener, C. D. (1979). Biogeography of the bees. *Annals of the Missouri Botanical Garden*, *66*(3), 277–347. doi:10.2307/2398833

Pedro, S. R. M., & Camargo, J. M. F. (2003). Meliponini neotropicais: O gênero Partamona Schwarz, 1939 (Hymenoptera, Apidae). *Revista Brasileira de Entomologia*, *47*, 1–117. doi:10.1590/S0085-56262003000500001

Revilla, G., Ramos, F. R., López-Nieto, M. J., Alvarez, E., & Martín, J. F. (1986). Glucose represses formation of delta-(L-alpha-aminoadipyl)-L-cysteinyl-D-valine and isopenicillin N synthase but not penicillin acyltransferase in Penicillium chrysogenum. *Journal of Bacteriology*, *168*(2), 947–952. doi:10.1128/jb.168.2.947-952.1986 PMID:3096965

Roubik, D. W. (1983). Nest and colony characteristics of stingless bees from Panama (Hymenoptera: Apidae). *Journal of the Kansas Entomological Society*, 327–355.

Roubik, D. W. (1992). *Ecology and natural history of tropical bees*. Cambridge University Press.

Roubik, D. W., Sakagami, S. F., & Kudo, I. (1985). A note on distribution and nesting of the Himalayan honey bee Apis laboriosa Smith (Hymenoptera: Apidae). *Journal of the Kansas Entomological Society*, 746–749.

Sakagami, S. F. (1975). Stingless bees (excl. Tetragonula) from the continental Southeast Asia in the collection of Berince P. Bishop Museum, Honolulu (Hymenoptera, Apidae)(with 14 text-figures and 3 tables). 北海道大學理學部紀要, *20*(1), 49–76.

Sakagami, S. F. (1978). Tetragonula Stingless bees of the continental Asia and Sri Lanka (Hymenoptera, Apidae)(with 124 text-figures, 1 plate and 36 tables). 北海道大學理學部紀要, *21*(2), 165–247.

Sakagami, S F, Inoue, T., & Salmah, S. (1990). Stingless bees of central Sumatra. *Stingless Bees of Central Sumatra*, 125–137.

Sakagami, S. F. (1961). Bees of Xylocopinae and Apidae collected by the Osaka City University Biological Expedetion to Southeast Asia 1957-58, with some biological notes. *Nature and Life SE Asia, 1*, 409–444.

Salmah, S. (1991). Incubation period in the Sumatra stingless bee Trigona (tetragonula) minangkabau. *Treubia*, *30*, 195–201.

Salmah, S. (1989). *Tempat dan volume beberapa jenis lebah yang terdapat di Sumatera (Hymenoptera: Apidae)*. Seminar Dan Kongres Biologi Nasional IX Di Padang.

Schwarz, H. F. (1948). Stingless bees (Meliponinae) of the western hemisphere. *Bulletin of the American Museum of Natural History*, *90*, 1–546.

Shimizu, K., & Morse, D. E. (2018). Silicatein: A unique silica-synthesizing catalytic triad hydrolase from marine sponge skeletons and its multiple applications. *Methods in Enzymology, 605*, 429–455. doi:10.1016/bs.mie.2018.02.025 PMID:29909834

Veth, P. J. (1881). *Midden-Sumatra: reizen en onderzoekingen der Sumatra-expeditie uitgerust door het Aardrijskundig genootschap, 1877-1879* (Vol. 1, Issue 1). Brill Archive.

Yoshikawa, K. (1969). Preliminary report on entomology of the Osaka City University 5th Scientific Expedition to Southeast Asia 1966-With descriptions of two new genera of stenogastrine wasps by J. van der Vecht. *Nature and Life in Southeast Asia, 6*, 153–182.

Yusop, S. A. T. W., Asaruddin, M. R., Sukairi, A. H., & Sabri, W. M. A. W. (2018). Cytotoxicity and Antimicrobial Activity of Propolis from Trigona itama Stingless Bees against Staphylococcus aureus and Escherichia coli. *Indonesian Journal of Pharmaceutical Science and Technology, 1*(1), 13–20.

Compilation of References

Abante, C. G. R. (2020). Mayon volcano cultural heritage as a source of place identity in Guinobatan, Albay, Philippines. *Bicol University R&D Journal*, *23*(1), 1–14.

Abd Jalil, M. A., Kasmuri, A. R., & Hadi, H. (2017). Stingless bee honey, the natural wound healer: A review. *Skin Pharmacology and Physiology*, *30*(2), 66–75. doi:10.1159/000458416 PMID:28291965

Abd Jamil, Z., & Zhari, I. (2016). Antibacterial and phenolic content of propolis produced by two Malaysian stingless bees, Heterotrigona itama and Geniotrigona thoracica. *International Journal of Pharmacognosy and Phytochemical Research*, *8*(1), 156–161.

Abd Razak, S. B., Aziz, A. A., Ali, N. A., Ali, M. F., & Visser, F. (2016). The sustainable integration of meliponiculture as an additional income stream for rubber smallholders in Malaysia. *Proceeding in CRI & IRRDB International Rubber Conference*, 21–22.

Abduh, M. Y., Adam, A., Fadhlullah, M., Putra, R. E., & Manurung, R. (2020). Production of propolis and honey from *Tetragonula laeviceps* cultivated in Modular Tetragonula Hives. *Heliyon*, *6*(11), 1–8. doi:10.1016/j.heliyon.2020.e05405 PMID:33204881

Abdullah, M., Bakhtan, M. A. H., & Mokhtar, S. A. (2017). Number Plate Recognition Of Malaysia Vehicles Using Smearing Algorithm. *Science International (Lahore)*, *29*(4), 823–827.

Adenekan, M. O., Amusa, N. A., Lawal, A. O., & Okpeze, V. E. (2010). Physico-chemical and microbiological properties of honey samples obtained from Ibadan. *Journal of Microbiology and Antimicrobials*, *2*(8), 100–104.

Agus, A., Agussalim, N., Umami, N., & Budisatria, I. G. S. (2019b). Evaluation of antioxidant activity, phenolic, flavonoid and Vitamin C content of several honeys produced by the Indonesian stingless bee: *Tetragonula laeviceps*. *Livestock Research for Rural Development*, *31*(10).

Agus, A., Agussalim, A., Umami, N., & Budisatria, I. G. S. (2019a). Effect of different beehives size and daily activity of stingless bee *Tetragonula laeviceps* on bee-pollen production. *Buletin Peternakan*, *43*(4), 242–246. doi:10.21059/buletinpeternak.v43i4.47865

Agussalim, A. A., Nurliyani, & Umami, N. (2019a). The sugar content profile of honey produced by the Indonesian stingless bee, *Tetragonula laeviceps*, from different regions. *Livestock Research for Rural Development*, *31*(6). http://www.lrrd.org/lrrd31/6/aguss31091.html

Agussalim, A. A., Nurliyani, & Umami, N. (2019b). Free amino acids profile of honey produced by the Indonesian stingless bee: *Tetragonula laeviceps. The 8th International Seminar on Tropical Animal Production*, 149–152.

Agussalim, A. A., Umami, N., & Budisatria, I. G. S. (2017). The Effect of Daily Activities Stingless Bees of Trigona sp. on Honey Production. *The 7th International Seminar on Tropical Animal Production*, 223–227.

Agussalim, N., Umami, N., & Agus, A. (2020). The honey and propolis production from Indonesian stingless bee: *Tetragonula laeviceps. Livestock Research for Rural Development, 32*(8).

Agussalim, U. N., & Erwan. (2015). Production of Stingless Bees (Trigona sp.) Propolis in Various Bee Hives Design. *The 6th International Seminar on Tropical Animal Production*, 335–338.

Agussalim, A., Agus, A., Nurliyani, Umami, N., & Budisatria, I. G. S. (2019). Physicochemical properties of honey produced by the Indonesian stingless bee: *Tetragonula laeviceps. IOP Conference Series. Earth and Environmental Science, 387*(1), 012084. Advance online publication. doi:10.1088/1755-1315/387/1/012084

Ahmed, S., & Othman, N. H. (2013). Review of the medicinal effects of tualang honey and a comparison with manuka honey. *The Malaysian Journal of Medical Sciences: MJMS, 20*(3), 6–13. PMID:23966819

Ajitha Nath, K. G. R., Jayakumaran Nair, A., & VS, S. (2018). *Comparison of proteins in two honey samples from Apis and stingless bee.* Academic Press.

Akharaiyi, F. C., & Lawal, H. A. (2016). Physicochemical analysis and mineral contents of honey from farmers in western states of Nigeria. *J. Nat. Sci. Res, 6*, 78–84.

Al-Hatamleh, M. A. I., Boer, J. C., Wilson, K. L., Plebanski, M., Mohamud, R., & Mustafa, M. Z. (2020). Antioxidant-based medicinal properties of stingless bee products: Recent progress and future directions. *Biomolecules, 10*(6), 1–28. doi:10.3390/biom10060923 PMID:32570769

Ali, M. F., Abdul Aziz, A., & Williams, A. (2020). Assessing yield and yield stability of hevea clones in the southern and central regions of Malaysia. *Agronomy (Basel), 10*(5), 1–15. doi:10.3390/agronomy10050643

Alimentarius, C. (2001). Revised Codex Standard for Honey. Codex Stan. 12-1981, Rev. 1 (1987). World Health Organization.

Alimentarius, C. (2001). Revised codex standard for honey. *Codex Stan, 12*, 1982.

Allameh, S. K., Daud, H., Yusoff, F. M., Saad, C. R., & Ideris, A. (2012). Isolation, identification and characterization of Leuconostoc mesenteroides as a new probiotic from intestine of snakehead fish (Channa striatus). *African Journal of Biotechnology, 11*(16), 3810–3816.

Alvarez, L. J., Reynaldi, F. J., Ramello, P. J., Garcia, M. L. G., Sguazza, G. H., Abrahamovich, A. H., & Lucia, M. (2018). Detection of honey bee viruses in Argentinian stingless bees (Hymenoptera: Apidae). *Insectes Sociaux, 65*(1), 191–197. doi:10.100700040-017-0587-2

Alvarez-Suarez, J. M., Tulipani, S., Díaz, D., Estevez, Y., Romandini, S., Giampieri, F., Damiani, E., Astolfi, P., Bompadre, S., & Battino, M. (2010). Antioxidant and antimicrobial capacity of several monofloral Cuban honeys and their correlation with color, polyphenol content and other chemical compounds. *Food and Chemical Toxicology*, *48*(8–9), 2490–2499. doi:10.1016/j.fct.2010.06.021 PMID:20558231

Andualem, B. (2014). Physico-chemical, microbiological and antibacterial properties of Apis mellipodae and Trigona spp. honey against bacterial pathogens. *World Journal of Agricultural Sciences*, *10*(3), 112–120.

Araújo, J. M. A. (2001). Food chemistry: Theory and practice. Federal University of Viçosa Press.

Armbruster, W. S. (1984). The role of resin in angiosperm pollination: Ecological and chemical considerations. *American Journal of Botany*, *71*(8), 1149–1160. doi:10.1002/j.1537-2197.1984.tb11968.x

Ávila, S., Hornung, P. S., Teixeira, G. L., Malunga, L. N., Apea-Bah, F. B., Beux, M. R., Beta, T., & Ribani, R. H. (2019). Bioactive compounds and biological properties of Brazilian stingless bee honey have a strong relationship with the pollen floral origin. *Food Research International*, *123*, 1–10. doi:10.1016/j.foodres.2019.01.068 PMID:31284956

Awang, N., Ali, N., Abd Majid, F. A., Hamzah, S., & Abd Razak, S. B. (2018). Total flavonoids and phenolic contents of sticky and hard propolis from 10 species of Indo-Malayan stingless bees. *The Malaysian Journal of Analytical Sciences*, *22*(5), 877–884.

Aween, M. M. (2012). Evaluation on antibacterial activity of Lactobacillus acidophilus strains isolated from honey. *American Journal of Applied Sciences*, *9*(6), 807–817. doi:10.3844/ajassp.2012.807.817

Aween, M. M., Hassan, Z., Muhialdin, B. J., Eljamel, Y. A., Al-Mabrok, A. S. W., & Lani, M. N. (2012). Antibacterial activity of lactobacillus acidophilus strains isolated from honey marketed in malaysia against selected multiple antibiotic resistant (mar) gram-positive bacteria. *Journal of Food Science*, *77*(7), 364–371. doi:10.1111/j.1750-3841.2012.02776.x PMID:22757710

Aween, M. M., Zaiton, H., & Belal, J. (2010). Antimicrobial Activity of Lactic Acid Bacteria Isolated from Honey. *Proceedings of the International Symposium on Lactic Acid Bacteria (ISLAB'10), University Putra Malaysia*, 25–27.

Azevedo, M. S., Seraglio, S. K. T., Rocha, G., Balderas, C. B., Piovezan, M., Gonzaga, L. V., de Barcellos Falkenberg, D., Fett, R., de Oliveira, M. A. L., & Costa, A. C. O. (2017). Free amino acid determination by GC-MS combined with a chemometric approach for geographical classification of bracatinga honeydew honey (Mimosa scabrella Bentham). *Food Control*, *78*, 383–392. doi:10.1016/j.foodcont.2017.03.008

Azmi, W. A., Samsuri, N., Hatta, M. F. M., Ghazi, R., & Seng, C. T. (2017). Effects of stingless bee (Heterotrigona itama) pollination on greenhouse cucumber (Cucumis sativus). *Malaysian Applied Biology*, *46*(1), 51–55.

Azmi, W. A., Seng, C. T., & Solihin, N. S. (2016). Pollination efficiency of the stingless bee, Heterotrigona itama (Hymenoptera: Apidae) on chili (Capsicum annuum) in greenhouse. *J. Trop. Plant Physiol*, *8*, 1–11.

Azmi, W. A., Wan Sembok, W. Z., Yusuf, N., Mohd. Hatta, M. F., Salleh, A. F., Hamzah, M. A. H., & Ramli, S. N. (2019). Effects of pollination by the indo-Malaya stingless bee (Hymenoptera: Apidae) on the quality of greenhouse-produced rockmelon. *Journal of Economic Entomology*, *112*(1), 20–24. doi:10.1093/jee/toy290 PMID:30277528

Azuma, H., Toyota, M., Asakawa, Y., Takaso, T., & Tobe, H. (2002). Floral scent chemistry of mangrove plants. *Journal of Plant Research*, *115*(1), 47–53. doi:10.1007102650200007 PMID:12884048

Bahri, S. (2018). Pollen profile by stingless bee (Heterotrigona itama) reared in rubber smallholding environment at Tepoh, Terengganu. *Malaysian Journal of Microscopy*, *14*(1).

Bai, Y., Dong, J.-J., Guan, D.-L., Xie, J.-Y., & Xu, S.-Q. (2016). Geographic variation in wing size and shape of the grasshopper Trilophidia annulata (Orthoptera: Oedipodidae): morphological trait variations follow an ecogeographical rule. *Scientific Reports*, *6*(1), 1–15. doi:10.1038rep32680 PMID:27597437

Baroga-Barbecho, J. B., & Cervancia, C. R. (2019). *Pest of Philippine stingless bees*. Philippine Entomologist.

Baroni, M. V., Chiabrando, G. A., Costa, C., & Wunderlin, D. A. (2002). Assessment of the floral origin of honey by SDS-page immunoblot techniques. *Journal of Agricultural and Food Chemistry*, *50*(6), 1362–1367. doi:10.1021/jf011214i PMID:11879003

Batista, E. C., Carvalho, L. R., Casarini, D. E., Carmona, A. K., Dos Santos, E. L., Da Silva, E. D., Dos Santos, R. A., Nakaie, C. R., Rojas, M. V. M., de Oliveira, S. M., Bader, M., D'Almeida, V., Martins, A. M., de Picoly Souza, K., & Pesquero, J. B. (2011). ACE activity is modulated by the enzyme α-galactosidase A. *Journal of Molecular Medicine*, *89*(1), 65–74. doi:10.100700109-010-0686-2 PMID:20941593

Batista, M. A., Ramalho, M., & Soares, A. E. E. (2003). Nesting sites and abundance of Meliponini (Hymenoptera: Apidae) in heterogeneous habitats of the Atlantic Rain Forest, Bahia, Brazil. *Lundiana: International Journal of Biodiversity*, *4*(1), 19–23.

Belay, A., Solomon, W. K., Bultossa, G., Adgaba, N., & Melaku, S. (2013). Physicochemical properties of the Harenna forest honey, Bale, Ethiopia. *Food Chemistry*, *141*(4), 3386–3392. doi:10.1016/j.foodchem.2013.06.035 PMID:23993497

Belina-Aldemita, M. D., Opper, C., Schreiner, M., & D'Amico, S. (2019). Nutritional composition of pot-pollen produced by stingless bees (Tetragonula biroi Friese) from the Philippines. *Journal of Food Composition and Analysis*, *82*, 103215. doi:10.1016/j.jfca.2019.04.003

Biesmeijer, J. C., & Ermers, M. C. W. (1999). Social foraging in stingless bees: How colonies of Melipona fasciata choose among nectar sources. *Behavioral Ecology and Sociobiology, 46*(2), 129–140. doi:10.1007002650050602

Biesmeijer, J. C., Giurfa, M., Koedam, D., Potts, S. G., Joel, D. M., & Dafni, A. (2005). Convergent evolution: Floral guides, stingless bee nest entrances, and insectivorous pitchers. *Naturwissenschaften, 92*(9), 444–450. doi:10.100700114-005-0017-6 PMID:16133103

Biluca, F. C., Braghini, F., Gonzaga, L. V., Costa, A. C. O., & Fett, R. (2016). Physicochemical profiles, minerals and bioactive compounds of stingless bee honey (Meliponinae). *Journal of Food Composition and Analysis, 50*, 61–69. doi:10.1016/j.jfca.2016.05.007

Blochtein, B., & Serrão, J. E. (2013). Seasonal production and spatial distribution of Melipona bicolor schencki (Apidae; Meliponini) castes in brood combs in southern Brazil. *Apidologie, 44*(2), 176–187. doi:10.100713592-012-0169-2

Blois, M. S. (1958). Antioxidant determinations by the use of a stable free radical. *Nature, 181*(4617), 1199–1200. doi:10.1038/1811199a0

Bocian, A., Buczkowicz, J., Jaromin, M., Hus, K. K., & Legáth, J. (2019). An effective method of isolating honey proteins. *Molecules (Basel, Switzerland), 24*(13), 2399. doi:10.3390/molecules24132399 PMID:31261846

Bogdanov, S. (1997). Nature and origin of the antibacterial substances in honey. *Lebensmittel-Wissenschaft + Technologie, 30*(7), 748–753. doi:10.1006/fstl.1997.0259

Bogdanov, S., Jurendic, T., Sieber, R., & Gallmann, P. (2008). Honey for nutrition and health: A review. *Journal of the American College of Nutrition, 27*(6), 677–689. doi:10.1080/0731572 4.2008.10719745 PMID:19155427

Bradford, M. M. (1976). A rapid and sensitive method for the quantitation of microgram quantities of protein utilizing the principle of protein-dye binding. *Analytical Biochemistry, 72*(1–2), 248–254. doi:10.1016/0003-2697(76)90527-3 PMID:942051

Brown, E., O'Brien, M., Georges, K., & Suepaul, S. (2020). Physical characteristics and antimicrobial properties of Apis mellifera, Frieseomelitta nigra and Melipona favosa bee honeys from apiaries in Trinidad and Tobago. *BMC Complementary Medicine and Therapies, 20*(1), 1–9. doi:10.118612906-020-2829-5 PMID:32178659

Brudzynski, K., & Sjaarda, C. (2014). Antibacterial compounds of Canadian honeys target bacterial cell wall inducing phenotype changes, growth inhibition and cell lysis that resemble action of β-lactam antibiotics. *PLoS One, 9*(9), 1–11. doi:10.1371/journal.pone.0106967 PMID:25191847

Brunet, J., & Sweet, H. R. (2006). Impact of insect pollinator group and floral display size on outcrossing rate. *Evolution; International Journal of Organic Evolution, 60*(2), 234–246. doi:10.1111/j.0014-3820.2006.tb01102.x PMID:16610316

Buchori, D., Rizali, A., Priawandiputra, W., Raffiudin, R., Sartiami, D., Pujiastuti, Y., Pradana, M. G., Meilin, A., Leatemia, J. A., & Sudiarta, I. P. (2022). Beekeeping and managed bee diversity in Indonesia: Perspective and preference of beekeepers. *Diversity (Basel)*, *14*(1), 1–14. doi:10.3390/d14010052

Bulgasem, B. Y., Hassan, Z., Huda-Faujan, N., Omar, R. H. A., Lani, M. N., & Alshelmani, M. I. (2017). Effect of pH, heat treatment and enzymes on the antifungal activity of lactic acid bacteria against Candida species. *Malaysian Journal of Microbiology*, *13*(3), 195–202. doi:10.21161/mjm.89416

Camargo, J. M. F. (1970). Nihos e biologia de algumas especies de Meliponideos (Hymenoptera: Apidae) de regiao de Porto Velho, Territorio de Rondonia, Brasil. *Revista de Biología Tropical*, *16*(2), 207–239.

Can, Z., Yildiz, O., Sahin, H., Turumtay, E. A., Silici, S., & Kolayli, S. (2015). An investigation of Turkish honeys: Their physico-chemical properties, antioxidant capacities and phenolic profiles. *Food Chemistry*, *180*, 133–141. doi:10.1016/j.foodchem.2015.02.024 PMID:25766810

Carpentier, S. C., Witters, E., Laukens, K., Deckers, P., Swennen, R., & Panis, B. (2005). Preparation of protein extracts from recalcitrant plant tissues: An evaluation of different methods for two-dimensional gel electrophoresis analysis. *Proteomics*, *5*(10), 2497–2507. doi:10.1002/pmic.200401222 PMID:15912556

Castaldo, S., & Capasso, F. (2002). Propolis, an old remedy used in modern medicine. *Fitoterapia*, *73*, S1–S6. doi:10.1016/S0367-326X(02)00185-5 PMID:12495704

Castro, M. L., Cury, J. A., Rosalen, P. L., Alencar, S. M., Ikegaki, M., Duarte, S., & Koo, H. (2007). Própolis do sudeste e nordeste do Brasil: Influência da sazonalidade na atividade antibacteriana e composição fenólica. *Quimica Nova*, *30*(7), 1512–1516. doi:10.1590/S0100-40422007000700003

Cauich, O., Quezada Euan, J. J. G., Ramírez, V. M., Valdovinos-Nuñez, G. R., & Moo-Valle, H. (2006). Pollination of habanero pepper (*Capsicum chinense*) and production in enclosures using the stingless bee *Nannotrigona perilampoides*. *Journal of Apicultural Research*, *45*(3), 125–130. doi:10.1080/00218839.2006.11101330

Cervancia, C. R. (2018). Management and conservation of Philippine bees. In *Asian beekeeping in the 21st Century* (pp. 307–321). Springer. doi:10.1007/978-981-10-8222-1_14

Chanchao, C. (2009). Antimicrobial activity by Trigona laeviceps (stingless bee) honey from Thailand. *Pakistan Journal of Medical Sciences*, *25*(3), 364–369.

Chanchao, C. (2013). Bioactivity of Honey and Propolis of Tetragonula laeviceps in Thailand. In *Pot-Honey* (pp. 495–505). Springer. doi:10.1007/978-1-4614-4960-7_36

Chinhinst, T. X., Grob, G. B. J., Meeuwsen, F. J. A. J., & Sommeijer, M. J. (2003). Patterns of male production in the stingless bee Melipona favosa (Apidae, Meliponini). *Apidologie*, *34*(2), 161–170. doi:10.1051/apido:2003008

Chin, N. L., & Sowndhararajan, K. (2020). A review on analytical methods for honey classification, identification and authentication. In *Honey Analysis-New Advances and Challenges*. IntechOpen. doi:10.5772/intechopen.90232

Cho, K. S., & Tony Ng, H. K. (2021). Tolerance intervals in statistical software and robustness under model misspecification. *Journal of Statistical Distributions and Applications*, 8(1), 1–49. doi:10.118640488-021-00123-2

Choudhari, M. K., Punekar, S. A., Ranade, R. V., & Paknikar, K. M. (2012). Antimicrobial activity of stingless bee (Trigona sp.) propolis used in the folk medicine of Western Maharashtra, India. *Journal of Ethnopharmacology*, 141(1), 363–367. doi:10.1016/j.jep.2012.02.047 PMID:22425711

Chuttong, B., Chanbang, Y., & Burgett, M. (2014). Meliponiculture: Stingless bee beekeeping in Thailand. *Bee World*, 91(2), 41–45. doi:10.1080/0005772X.2014.11417595

Chuttong, B., Chanbang, Y., Sringarm, K., & Burgett, M. (2016). Physicochemical profiles of stingless bee (Apidae: Meliponini) honey from South east Asia (Thailand). *Food Chemistry*, 192, 149–155. doi:10.1016/j.foodchem.2015.06.089 PMID:26304332

Cicco, N., & Lattanzio, V. (2011). The Influence of Initial Carbonate Concentration on the Folin-Ciocalteu Micro-Method for the Determination of Phenolics with Low Concentration in the Presence of Me-thanol: A Comparative Study of Real-Time Monitored Reactions. *American Journal of Analytical Chemistry*, 2(7), 840–848. doi:10.4236/ajac.2011.27096

Ciulu, M., Farre, R., Floris, I., Nurchi, V. M., Panzanelli, A., Pilo, M. I., Spano, N., & Sanna, G. (2013). Determination of 5-hydroxymethyl-2-furaldehyde in royal jelly by a rapid reversed phase HPLC method. *Analytical Methods*, 5(19), 5010–5013. doi:10.1039/c3ay40634b

Contrera, F. A. L., Menezes, C., & Venturieri, G. C. (2011). *New horizons on stingless beekeeping (Apidae, Meliponini)*. Academic Press.

Corradini, E., Schmidt, P. J., Meynard, D., Garuti, C., Montosi, G., Chen, S., Vukicevic, S., Pietrangelo, A., Lin, H. Y., & Babitt, J. L. (2010). BMP6 treatment compensates for the molecular defect and ameliorates hemochromatosis in Hfe knockout mice. *Gastroenterology*, 139(5), 1721–1729. doi:10.1053/j.gastro.2010.07.044 PMID:20682319

Cortopassi-Laurino, M., Imperatriz-Fonseca, V. L., Roubik, D. W., Dollin, A., Heard, T., Aguilar, I., Venturieri, G. C., Eardley, C., & Nogueira-Neto, P. (2006). Global meliponiculture: Challenges and opportunities. *Apidologie*, 37(2), 275–292. doi:10.1051/apido:2006027

Couvillon, M. J., Wenseleers, T., Imperatriz-Fonseca, V. L., Nogueira-Neto, P., & Ratnieks, F. L. W. (2008). Comparative study in stingless bees (Meliponini) demonstrates that nest entrance size predicts traffic and defensivity. *Journal of Evolutionary Biology*, 21(1), 194–201. doi:10.1111/j.1420-9101.2007.01457.x PMID:18021200

Cruz, D. de O., Freitas, B. M., da Silva, L. A., da Silva, E. M. S., & Bomfim, I. G. A. (2005). Pollination efficiency of the stingless bee Melipona subnitida on greenhouse sweet pepper. *Pesquisa Agropecuária Brasileira*, 40(12), 1197–1201. doi:10.1590/S0100-204X2005001200006

Cui, Z.-W., Sun, L.-J., Chen, W., & Sun, D.-W. (2008). Preparation of dry honey by microwave–vacuum drying. *Journal of Food Engineering, 84*(4), 582–590. doi:10.1016/j.jfoodeng.2007.06.027

Cunha, I., Sawaya, A. C. H. F., Caetano, F. M., Shimizu, M. T., Marcucci, M. C., Drezza, F. T., Povia, G. S., & Carvalho, P. de O. (2004). Factors that influence the yield and composition of Brazilian propolis extracts. *Journal of the Brazilian Chemical Society, 15*(6), 964–970. doi:10.1590/S0103-50532004000600026

da Silva, I. A. A., da Silva, T. M. S., Camara, C. A., Queiroz, N., Magnani, M., de Novais, J. S., Soledade, L. E. B., de Oliveira Lima, E., de Souza, A. L., & de Souza, A. G. (2013). Phenolic profile, antioxidant activity and palynological analysis of stingless bee honey from Amazonas, Northern Brazil. *Food Chemistry, 141*(4), 3552–3558. doi:10.1016/j.foodchem.2013.06.072 PMID:23993520

da Silva, P. M., Gauche, C., Gonzaga, L. V., Costa, A. C. O., & Fett, R. (2016). Honey: Chemical composition, stability and authenticity. *Food Chemistry, 196*, 309–323. doi:10.1016/j.foodchem.2015.09.051 PMID:26593496

da Veiga, J. E. (2012). *O desenvolvimento agrícola: uma visão histórica.* Edusp.

Dajanta, K., Chukeatirote, E., & Apichartsrangkoon, A. (2012). Nutritional and physicochemical qualities of Thua Nao (Thai traditional fermented soybean). *Warasan Khana Witthayasat Maha Witthayalai Chiang Mai, 39*(4), 562–574.

de Almeida-Muradian, L. B., Stramm, K. M., Horita, A., Barth, O. M., da Silva de Freitas, A., & Estevinho, L. M. (2013). Comparative study of the physicochemical and palynological characteristics of honey from M elipona subnitida and A pis mellifera. *International Journal of Food Science & Technology, 48*(8), 1698–1706. doi:10.1111/ijfs.12140

de Araújo Filho, J. R. (1957). A cultura da banana no Brasil. *Boletim Paulista de Geografia, 27*, 27–54.

de Oliveira, E. M., & Oliveira, F. L. C. (2018). Forecasting mid-long term electric energy consumption through bagging ARIMA and exponential smoothing methods. *Energy, 144*, 776–788. doi:10.1016/j.energy.2017.12.049

de Sousa, J. M. B., de Souza, E. L., Marques, G., de Toledo Benassi, M., Gullón, B., Pintado, M. M., & Magnani, M. (2016). Sugar profile, physicochemical and sensory aspects of monofloral honeys produced by different stingless bee species in Brazilian semi-arid region. *Lebensmittel-Wissenschaft + Technologie, 65*, 645–651. doi:10.1016/j.lwt.2015.08.058

Del Sarto, M. C. L., Peruquetti, R. C., & Campos, L. A. O. (2005). Evaluation of the neotropical stingless bee Melipona quadrifasciata (Hymenoptera: Apidae) as pollinator of greenhouse tomatoes. *Journal of Economic Entomology, 98*(2), 260–266. doi:10.1093/jee/98.2.260 PMID:15889711

Di Girolamo, F., D'Amato, A., & Righetti, P. G. (2012). Assessment of the floral origin of honey via proteomic tools. *Journal of Proteomics, 75*(12), 3688–3693. doi:10.1016/j.jprot.2012.04.029 PMID:22571915 do Nascimento, D. L., & Nascimento, F. S. (2012). Extreme effects of season on the foraging activities and colony productivity of a stingless bee (Melipona asilvai Moure, 1971) in Northeast Brazil. *Psyche*, 2012.

Dodologlu, A., Emsen, B., & Gene, F. (2004). Comparison of some characteristics of queen honey bees (Apis mellifera L.) reared by using Doolittle method and natural queen cells. *Journal of Applied Animal Research*, *26*(2), 113–115. doi:10.1080/09712119.2004.9706518

Dollin, A. E., Dollin, L. J., & Sakagami, S. F. (1997). Australian stingless bees of the genus Trigona (Hymenoptera: Apidae). *Invertebrate Systematics*, *11*(6), 861–896. doi:10.1071/IT96020

dos Santos, S. A. B., Roselino, A. C., & Bego, L. R. (2008). Pollination of cucumber, Cucumis sativus L.(Cucurbitales: Cucurbitaceae), by the stingless bees Scaptotrigona aff. depilis moure and Nannotrigona testaceicornis Lepeletier (Hymenoptera: Meliponini) in greenhouses. *Neotropical Entomology*, *37*(5), 506–512. doi:10.1590/S1519-566X2008000500002 PMID:19061034

Duell, M. E. (2018). *Matters of size: behavioral, morphological, and physiological performance scaling among stingless bees (Meliponini)*. Arizona State University.

Dunne, C., O'Mahony, L., Murphy, L., Thornton, G., Morrissey, D., O'Halloran, S., Feeney, M., Flynn, S., Fitzgerald, G., Daly, C., Kiely, B., O'Sullivan, G. C., Shanahan, F., & Collins, J. K. (2001). In vitro selection criteria for probiotic bacteria of human origin: Correlation with in vivo findings. *The American Journal of Clinical Nutrition*, *73*(2), 386s–392s. doi:10.1093/ajcn/73.2.386s PMID:11157346

El Sohaimy, S. A., Masry, S. H. D., & Shehata, M. G. (2015). Physicochemical characteristics of honey from different origins. *Annals of Agricultural Science*, *60*(2), 279–287. doi:10.1016/j.aoas.2015.10.015

Eltz, T., Brühl, C. A., Van der Kaars, S., & Linsenmair, E. K. (2002). Determinants of stingless bee nest density in lowland dipterocarp forests of Sabah, Malaysia. *Oecologia*, *131*(1), 27–34. doi:10.100700442-001-0848-6 PMID:28547507

Erwan, A., M., S., Muhsinin, M., & Agussalim. (2020). The effect of different beehives on the activity of foragers, honey potsnumber and honey production from stingless bee Tetragonula sp. *Livestock Research for Rural Development, 32*(10).

Erwan, S., Syamsuhaidi, P., D. K., Muhsinin, M., & Agussalim. (2021). *Propolis mixture production and foragers daily activity of stingless bee Tetragonula sp . in bamboo and box hives.* Academic Press.

Escuredo, O., Dobre, I., Fernández-González, M., & Seijo, M. C. (2014). Contribution of botanical origin and sugar composition of honeys on the crystallization phenomenon. *Food Chemistry*, *149*, 84–90. doi:10.1016/j.foodchem.2013.10.097 PMID:24295680

Escuredo, O., Míguez, M., Fernández-González, M., & Carmen Seijo, M. (2013). Nutritional value and antioxidant activity of honeys produced in a European Atlantic area. *Food Chemistry*, *138*(2-3), 851–856. doi:10.1016/j.foodchem.2012.11.015 PMID:23411187

Escuredo, O., Silva, L. R., Valentão, P., Seijo, M. C., & Andrade, P. B. (2012). Assessing Rubus honey value: Pollen and phenolic compounds content and antibacterial capacity. *Food Chemistry*, *130*(3), 671–678. doi:10.1016/j.foodchem.2011.07.107

Fadzli, M. H., Jaafar, T. N. A. M., Ali, M. S., Nur, N. F. M., Tan, M. P., & Piah, R. M. (2022). Reproductive aspects of the coastal trevally, Carangoides coeruleopinnatus in Terengganu Waters, Malaysia. *Aquaculture and Fisheries*, *7*(5), 500–506. doi:10.1016/j.aaf.2022.04.009

Fatima, I. J., AB, M. H., Salwani, I., & Lavaniya, M. (2018). Physicochemical characteristics of malaysian stingless bee honey from trigona species. *IIUM Medical Journal Malaysia*, *17*(1), 187–191. doi:10.31436/imjm.v17i1.1030

Finola, M. S., Lasagno, M. C., & Marioli, J. M. (2007). Microbiological and chemical characterization of honeys from central Argentina. *Food Chemistry*, *100*(4), 1649–1653. doi:10.1016/j.foodchem.2005.12.046

Firdaus, A., Khalid, J., & Yong, Y. K. (2018). Malaysian Tualang honey and its potential anti-cancer properties: A review. *Sains Malaysiana*, *47*(11), 2705–2711. doi:10.17576/jsm-2018-4711-14

Fonseca, V. L. I. (2012). *Best management practices in agriculture for sustainable use and conservation of pollinators*. Academic Press.

Gallai, N., Salles, J.-M., Settele, J., & Vaissière, B. E. (2009). Economic valuation of the vulnerability of world agriculture confronted with pollinator decline. *Ecological Economics*, *68*(3), 810–821. doi:10.1016/j.ecolecon.2008.06.014

Gela, A., Hora, Z. A., Kebebe, D., & Gebresilassie, A. (2021). Physico-chemical characteristics of honey produced by stingless bees (Meliponula beccarii) from West Showa zone of Oromia Region, Ethiopia. *Heliyon*, *7*(1), 1–7. doi:10.1016/j.heliyon.2020.e05875 PMID:33506124

Ghatge, A. L. (1951). The bees of India. *Indian Bee J.*, *13*, 88.

Gilman, E. F. (1999). Ixora coccinea. *University of Florida, Cooperative Extension Service. Institute of Food and Agriculture Sciences. Fact Sheet, FPS-291*, 1–3.

Goff, R. C. (1976). Vertical structure of thunderstorm outflows. *Monthly Weather Review*, *104*(11), 1429–1440. doi:10.1175/1520-0493(1976)104<1429:VSOTO>2.0.CO;2

Griffith, G., & Watson, A. (2016). Agricultural markets and marketing policies. *The Australian Journal of Agricultural and Resource Economics*, *60*(4), 594–609. doi:10.1111/1467-8489.12161

Grüter, C. (2020). *Stingless bees*. Springer International Publishing. doi:10.1007/978-3-030-60090-7

Guerrini, A., Bruni, R., Maietti, S., Poli, F., Rossi, D., Paganetto, G., Muzzoli, M., Scalvenzi, L., & Sacchetti, G. (2009). Ecuadorian stingless bee (Meliponinae) honey: A chemical and functional profile of an ancient health product. *Food Chemistry*, *114*(4), 1413–1420. doi:10.1016/j.foodchem.2008.11.023

Halawani, E. M., & Shohayeb, M. M. (2011). Shaoka and Sidr honeys surpass in their antibacterial activity local and imported honeys available in Saudi markets against pathogenic and food spoilage bacteria. *Australian Journal of Basic and Applied Sciences*, *5*(4), 187–191.

Hamid, S. A., Salleh, M. S., Thevan, K., & Hashim, N. A. (2016). Distribution and morphometrical variations of stingless bees (Apidae: Meliponini) in urban and forest areas of Penang Island, Malaysia. *J. Trop. Resour. Sustain. Sci*, *4*, 1–5.

Hartfelder, K., & Engels, W. (1989). The composition of larval food in stingless bees: Evaluating nutritional balance by chemosystematic methods. *Insectes Sociaux*, *36*(1), 1–14. doi:10.1007/BF02225876

Hartfelder, K., Makert, G. R., Judice, C. C., Pereira, G. A. G., Santana, W. C., Dallacqua, R., & Bitondi, M. M. G. (2006). Physiological and genetic mechanisms underlying caste development, reproduction and division of labor in stingless bees. *Apidologie*, *37*(2), 144–163. doi:10.1051/apido:2006013

Hasali, N. H. M., Zamri, A. I., Lani, M. N., Mubarak, A., & Suhaili, Z. (2015). Identification of lactic acid bacteria from Meliponine honey and their antimicrobial activity against pathogenic bacteria. *American-Eurasian Journal of Sustainable Agriculture*, *9*(6), 1–7.

Hasali, N. O. R. H., Zamri, A. I., Lani, M. N., Mubarak, A., Ahmad, F., & Chilek, T. Z. T. (2018). Physico-chemical analysis and antibacterial activity of raw honey of stingless bee farmed in coastal areas in Kelantan and Terengganu. *Malaysian Applied Biology*, *47*(4), 145–151.

Hasan, A. E. Z., Nashrianto, H., Juhaeni, R. N., & Artika, I. M. (2016). Optimization of conditions for flavonoids extraction from mangosteen (Garcinia mangostana L.). *Der Pharmacia Lettre*, *8*(18), 114–120.

Hasnol, N. D. S., Jinap, S., & Sanny, M. (2014). Effect of different types of sugars in a marinating formulation on the formation of heterocyclic amines in grilled chicken. *Food Chemistry*, *145*, 514–521. doi:10.1016/j.foodchem.2013.08.086 PMID:24128508

Heard, T. A. (1999). The role of stingless bees in crop pollination. *Annual Review of Entomology*, *44*(1), 183–206. doi:10.1146/annurev.ento.44.1.183 PMID:15012371

Herwina, H., Salmah, S., Janra, M. N., Nurdin, J., & Sari, D. A. (2021). Stingless beekeeping (Hymenoptera: Apidae: Meliponini) and Its Potency for Other Related-Ventures in West Sumatra. *Journal of Physics: Conference Series*, *1940*(1), 12073. doi:10.1088/1742-6596/1940/1/012073

Hoque, M. Z., Akter, F., Hossain, K. M., Rahman, M. S. M., Billah, M. M., & Islam, K. M. D. (2010). Isolation, identification and analysis of probiotic properties of Lactobacillus spp. from selective regional yoghurts. *World J Dairy Food Sci*, *5*(1), 39–46.

Ibáñez, A., Riveros, R., Hurtado, E., Gleichgerrcht, E., Urquina, H., Herrera, E., Amoruso, L., Reyes, M. M., & Manes, F. (2012). The face and its emotion: Right N170 deficits in structural processing and early emotional discrimination in schizophrenic patients and relatives. *Psychiatry Research*, *195*(1–2), 18–26. doi:10.1016/j.psychres.2011.07.027 PMID:21824666

Ibarguren, C., Raya, R. R., Apella, M. C., & Audisio, M. C. (2010). Enterococcus faecium isolated from honey synthesized bacteriocin-like substances active against different Listeria monocytogenes strains. *Journal of Microbiology (Seoul, Korea)*, *48*(1), 44–52. doi:10.100712275-009-0177-8 PMID:20221729

Inoue, T., & Salmah, S. (1990). Nest site selection and reproductive ecology of the Asian honey bee, Apis cerana indica, in central Sumatra. *Nest Site Selection and Reproductive Ecology of the Asian Honey Bee, Apis Cerana Indica, in Central Sumatra.*, 219–232.

Jamnik, P., Raspor, P., & Javornik, B. (2012). A proteomic approach for investigation of bee products: Royal jelly, propolis and honey. *Food Technology and Biotechnology*, *50*(3), 270–274.

Jibril, F. I., Hilmi, A. B. M., & Manivannan, L. (2019). Isolation and characterization of polyphenols in natural honey for the treatment of human diseases. *Bulletin of the National Research Center*, *43*(1), 1–9. doi:10.118642269-019-0044-7

Jongjitvimol, T., & Wattanachaiyingcharoen, W. (2006). Pollen food sources of the stingless bees Trigona apicalis Smith, 1857, Trigona collina Smith, 1857 and Trigona fimbriata Smith, 1857 (Apidae, Meliponinae) in Thailand. *Tropical Natural History*, *6*(2), 75–82.

Jo, W. S., Yang, K. M., Park, H. S., Kim, G. Y., Nam, B. H., Jeong, M. H., & Choi, Y. J. (2012). Effect of microalgal extracts of Tetraselmis suecica against UVB-induced photoaging in human skin fibroblasts. *Toxicological Research*, *28*(4), 241–248. doi:10.5487/TR.2012.28.4.241 PMID:24278616

Kahono, S., Chantawannakul, P., & Engel, M. S. (2018). Social bees and the current status of beekeeping in Indonesia. In *Asian beekeeping in the 21st century* (pp. 287–306). Springer. doi:10.1007/978-981-10-8222-1_13

Karabagias, I. K., Badeka, A., Kontakos, S., Karabournioti, S., & Kontominas, M. G. (2014). Characterisation and classification of Greek pine honeys according to their geographical origin based on volatiles, physicochemical parameters and chemometrics. *Food Chemistry*, *146*, 548–557. doi:10.1016/j.foodchem.2013.09.105 PMID:24176380

Kelly, N., Farisya, M. S. N., Kumara, T. K., & Marcela, P. (2014). Species Diversity and External Nest Characteristics of Stingless Bees in Meliponiculture. *Pertanika. Journal of Tropical Agricultural Science*, *37*(3), 293–298.

Kerr, W. E., Carvalho, G. A., Silva, A. C., da, & Assis, M. da G. P. de. (2010). Aspectos pouco mencionados da biodiversidade amazônica. *Parcerias Estratégicas*, *6*(12), 20–41.

Khalifa, S. A. M., Elashal, M. H., Yosri, N., Du, M., Musharraf, S. G., Nahar, L., Sarker, S. D., Guo, Z., Cao, W., Zou, X., Abd El-Wahed, A. A., Xiao, J., Omar, H. A., Hegazy, M.-E. F., & El-Seedi, H. R. (2021). Bee pollen: Current status and therapeutic potential. *Nutrients*, *13*(6), 1876. doi:10.3390/nu13061876 PMID:34072636

Khongkwanmueang, A., Nuyu, A., Straub, L., & Maitip, J. (2020). Physicochemical profiles, antioxidant and antibacterial capacity of honey from stingless bee tetragonula laeviceps species complex. *E3S Web of Conferences, 141*, 3007.

Kiros, W., & Tsegay, T. (2017). Honey-bee production practices and hive technology preferences in Jimma and Illubabor Zone of Oromiya Regional State, Ethiopia. *Agriculture and Environment, 9*(1), 31–43.

Klaenhammer, T. R. (1993). Genetics of bacteriocins produced by lactic acid bacteria. *FEMS Microbiology Reviews, 12*(1–3), 39–85. doi:10.1016/0168-6445(93)90057-G PMID:8398217

Klakasikorn, A., Wongsiri, S., Deowanish, S., & Duangphakdee, O. (2005). New record of stingless bees (Meliponini: Trigona) in Thailand. *Tropical Natural History, 5*(1), 1–7.

Klingenberg, C. P. (2002). Morphometrics and the role of the phenotype in studies of the evolution of developmental mechanisms. *Gene, 287*(1–2), 3–10. doi:10.1016/S0378-1119(01)00867-8 PMID:11992717

Koedam, D., Aponte, O. I. C., & Imperatriz-Fonseca, V. L. (2007). Egg laying and oophagy by reproductive workers in the polygynous stingless bee Melipona bicolor (Hymenoptera, Meliponini). *Apidologie, 38*(1), 55–66. doi:10.1051/apido:2006053

Koeniger, N., & Koeniger, G. (1980). Observations and experiments on migration and dance communication of Apsis dorsata in Sri Lanka. *Journal of Apicultural Research, 19*(1), 21–34. doi:10.1080/00218839.1980.11099994

Kruger, N. J. (2002). Detection of polypeptides on immunoblots using enzyme-conjugated or radiolabeled secondary ligands. In *The Protein Protocols Handbook* (pp. 405–414). Springer. doi:10.1385/1-59259-169-8:405

Kuberappa, G. C., Mohite, S., & Kencharaddi, R. N. (2005). Biometrical variations among populations of stingless bee, Trigona iridipennis in Karnataka. *Indian Bee Journal, 67*, 145–149.

Kumazawa, S., Hamasaka, T., & Nakayama, T. (2004). Antioxidant activity of propolis of various geographic origins. *Food Chemistry, 84*(3), 329–339. doi:10.1016/S0308-8146(03)00216-4

Kwakman, P. H. S., te Velde, A. A., de Boer, L., Speijer, D., Christina Vandenbroucke-Grauls, M. J., & Zaat, S. A. J. (2010). How honey kills bacteria. *The FASEB Journal, 24*(7), 2576–2582. doi:10.1096/fj.09-150789 PMID:20228250

Laallam, H., Boughediri, L., Bissati, S., Menasria, T., Mouzaoui, M. S., Hadjadj, S., Hammoudi, R., & Chenchouni, H. (2015). Modeling the synergistic antibacterial effects of honey characteristics of different botanical origins from the Sahara Desert of Algeria. *Frontiers in Microbiology, 6*, 1–12. doi:10.3389/fmicb.2015.01239 PMID:26594206

Laemmli, U. K. (1970). Cleavage of structural proteins during the assembly of the head of bacteriophage T4. *Nature, 227*(5259), 680–685. doi:10.1038/227680a0 PMID:5432063

Laksono, P., Raffiudin, R., & Juliandi, B. (2020). Stingless bee Tetragonula laeviceps and T. aff. biroi: Geometric morphometry analysis of wing venation variations. *IOP Conference Series. Earth and Environmental Science, 457*(1), 12084. doi:10.1088/1755-1315/457/1/012084

Lancaster, L. T., Dudaniec, R. Y., Chauhan, P., Wellenreuther, M., Svensson, E. I., & Hansson, B. (2016). Gene expression under thermal stress varies across a geographical range expansion front. *Molecular Ecology, 25*(5), 1141–1156. doi:10.1111/mec.13548 PMID:26821170

Lani, M. N., Zainudin, A. H., Razak, S. B. A., Mansor, A., & Hassan, Z. (2017). Microbiological quality and pH changes of honey produced by stingless bees, Heterotrigona itama and Geniotrigona thoracica stored at ambient temperature. *Malaysian Applied Biology, 46*(3), 89–96.

Lee, D.-C., Lee, S.-Y., Cha, S.-H., Choi, Y.-S., & Rhee, H.-I. (1998). Discrimination of native bee-honey and foreign bee-honey by SDS-PAGE. *Korean Journal of Food Science Technology, 30*(1), 1–5. doi:10.9721/KJFST.2017.49.1.1

Lee, H., Churey, J. J., & Worobo, R. W. (2008). Antimicrobial activity of bacterial isolates from different floral sources of honey. *International Journal of Food Microbiology, 126*(1–2), 240–244. doi:10.1016/j.ijfoodmicro.2008.04.030 PMID:18538876

Lob, S., Afiffi, N., Razak, S. B. A., Ibrahim, N. F., & Nawi, I. H. M. (2017). Composition and identification of pollen collected by stingless bee (Heterotrigona itama) in forested and coastal area of Terengganu. *Malaysian Applied Biology Journal, 46*(3), 227–232.

Lusby, D. A. (1996). Small cell size foundation for mite control. *American Bee Journal, 136*(7), 468–470.

Lusby, P. E., Coombes, A. L., & Wilkinson, J. M. (2005). Bactericidal activity of different honeys against pathogenic bacteria. *Archives of Medical Research, 36*(5), 464–467. doi:10.1016/j.arcmed.2005.03.038 PMID:16099322

Maa, T.-C. (1953). *An inquiry into the systematics of the tribus Apidini or honeybees (Hym.).* Archipel.

Maitip, J., Mookhploy, W., Khorndork, S., & Chantawannakul, P. (2021). Comparative Study of Antimicrobial Properties of Bee Venom Extracts and Melittins of Honey Bees. *Antibiotics (Basel, Switzerland), 10*(12), 1–14. doi:10.3390/antibiotics10121503 PMID:34943715

Marcucci, M. C. (1995). Propolis: Chemical composition, biological properties and therapeutic activity. *Apidologie, 26*(2), 83–99. doi:10.1051/apido:19950202

Margono, B. A., Potapov, P. V., Turubanova, S., Stolle, F., & Hansen, M. C. (2014). Primary forest cover loss in Indonesia over 2000–2012. *Nature Climate Change, 4*(8), 730–735. doi:10.1038/nclimate2277

Marshall, T., & Williams, K. M. (1987). Electrophoresis of honey: Characterization of trace proteins from a complex biological matrix by silver staining. *Analytical Biochemistry, 167*(2), 301–303. doi:10.1016/0003-2697(87)90168-0 PMID:2450485

McGregor, S. E. (1976). Insect pollination of cultivated crop plants (Issue 496). Agricultural Research Service, US Department of Agriculture.

McLoone, P., Warnock, M., & Fyfe, L. (2016). Honey: A realistic antimicrobial for disorders of the skin. *Journal of Microbiology, Immunology, and Infection, 49*(2), 161–167. doi:10.1016/j.jmii.2015.01.009 PMID:25732699

Medic-Saric, M., Bojic, M., Rastija, V., & Cvek, J. (2013). Polyphenolic profiling of Croatian propolis and wine. *Food Technology and Biotechnology, 51*(2), 159–170.

Mercan, N. (2006). Antimicrobial activity and chemical compositions of Turkish propolis from different regions. *African Journal of Biotechnology, 5*(11), 1151–1153.

Michener, C. D. (1979). Biogeography of the bees. *Annals of the Missouri Botanical Garden, 66*(3), 277–347. doi:10.2307/2398833

Michener, C. D. (2007). *The bees of the world* (2nd ed.). The Johns Hopkins University Press.

Michener, C. D. (2013). The meliponini. In *Pot-honey* (pp. 3–17). Springer. doi:10.1007/978-1-4614-4960-7_1

Michener, C. D., & Michener, C. D. (1974). *The social behavior of the bees: a comparative study*. Harvard University Press.

Mokhtar, S. U. (2019). Comparison of total phenolic and flavonoids contents in Malaysian propolis extract with two different extraction solvents. *International Journal of Engineering Technology and Sciences, 6*(2), 1–11. doi:10.15282/ijets.v6i2.2577

Molan, P. C. (1992). The antibacterial activity of honey: 1. The nature of the antibacterial activity. *Bee World, 73*(1), 5–28. doi:10.1080/0005772X.1992.11099109

Monthly Manufacturing Statistics Malaysia. (2014). Department of Statistics.

Moo-Huchin, V. M., González-Aguilar, G. A., Lira-Maas, J. D., Pérez-Pacheco, E., Estrada-León, R., Moo-Huchin, M. I., & Sauri-Duch, E. (2015). Physicochemical properties of Melipona beecheii honey of the Yucatan Peninsula. *Journal of Food Research, 4*(5), 25–32. doi:10.5539/jfr.v4n5p25

Naila, A., Flint, S. H., Sulaiman, A. Z., Ajit, A., & Weeds, Z. (2018). Classical and novel approaches to the analysis of honey and detection of adulterants. *Food Control, 90*, 152–165. doi:10.1016/j.foodcont.2018.02.027

Ndubisi, N. O., Malhotra, N. K., & Wah, C. K. (2008). Relationship marketing, customer satisfaction and loyalty: A theoretical and empirical analysis from an Asian perspective. *Journal of International Consumer Marketing, 21*(1), 5–16. doi:10.1080/08961530802125134

Ng, W.-J., Sit, N.-W., Ooi, P. A.-C., Ee, K.-Y., & Lim, T.-M. (2020). The antibacterial potential of honeydew honey produced by stingless bee (Heterotrigona itama) against antibiotic resistant bacteria. *Antibiotics (Basel, Switzerland), 9*(12), 1–16. doi:10.3390/antibiotics9120871 PMID:33291356

Nicodemo, D., Malheiros, E. B., De Jong, D., & Nogueira Couto, R. H. (2013). Enhanced production of parthenocarpic cucumbers pollinated with stingless bees and Africanized honey bees in greenhouses. *Semina. Ciências Agrárias*, *34*(6Supl1), 3625–3633. doi:10.5433/1679-0359.2013v34n6Supl1p3625

Nogueira-Neto, P. (1997). *Vida e criação de abelhas indígenas sem ferrão*. Academic Press.

Nolan, V. C., Harrison, J., & Cox, J. A. G. (2019). Dissecting the antimicrobial composition of honey. *Antibiotics (Basel, Switzerland)*, *8*(4), 1–26. doi:10.3390/antibiotics8040251 PMID:31817375

Nombré, I., Schweitzer, P., Boussim, J. I., & Rasolodimby, J. M. (2010). Impacts of storage conditions on physicochemical characteristics of honey samples from Burkina Faso. *African Journal of Food Science*, *4*(7), 458–463.

Nordin, A., Sainik, N. Q. A. V., Chowdhury, S. R., Saim, A., & Idrus, R. B. H. (2018). Physicochemical properties of stingless bee honey from around the globe: A comprehensive review. *Journal of Food Composition and Analysis*, *73*, 91–102. doi:10.1016/j.jfca.2018.06.002

Norowi, M. H., Mohd, F., Sajap, A. S., Rosliza, J., & Suri, R. (2010). Conservation and sustainable utilization of stingless bees for pollination services in agricultural ecosystems in Malaysia. *Proceedings of International Seminar on Enhancement of Functional Biodiversity Relevant to Sustainable Food Production in ASPAC*, 1–11.

Nunes-Silva, P., Hrncir, M., Shipp, L., Kevan, P., & Imperatriz-Fonseca, V. L. (2013). The behaviour of Bombus impatiens (Apidae, Bombini) on tomato (Lycopersicon esculentum Mill., Solanaceae) flowers: Pollination and reward perception. *Journal of Pollination Ecology*, *11*, 33–40. doi:10.26786/1920-7603(2013)3

Oddo, L. P., Heard, T. A., Rodríguez-Malaver, A., Pérez, R. A., Fernández-Muiño, M., Sancho, M. T., Sesta, G., Lusco, L., & Vit, P. (2008). Composition and antioxidant activity of Trigona carbonaria honey from Australia. *Journal of Medicinal Food*, *11*(4), 789–794. doi:10.1089/jmf.2007.0724 PMID:19012514

Oldroyd, B. P., & Wongsiri, S. (2009). *Asian honey bees: biology, conservation, and human interactions*. Harvard University Press. doi:10.2307/j.ctv2drhcfb

Oleksa, A., & Tofilski, A. (2015). Wing geometric morphometrics and microsatellite analysis provide similar discrimination of honey bee subspecies. *Apidologie*, *46*(1), 49–60. doi:10.100713592-014-0300-7

Oliveira, L. K. de. (2012). *Energia como Recurso de Poder na Política Internacional: geopolítica, estratégia e o papel do Centro de Decisão Energética*. Academic Press.

Oliveira, R. N., Mancini, M. C., Oliveira, F. C. S., de, Passos, T. M., Quilty, B., & Thiré, R. M. (2016). FTIR analysis and quantification of phenols and flavonoids of five commercially available plants extracts used in wound healing. *Matéria (Rio de Janeiro)*, *21*(3), 767–779. doi:10.1590/S1517-707620160003.0072

Omar, S., Mat-Kamir, N. F., & Sanny, M. (2019). Antibacterial activity of Malaysian produced stingless-bee honey on wound pathogens. *Journal of Sustainability Science and Management*, *14*, 67–79.

Özbalci, B., Boyaci, İ. H., Topcu, A., Kadılar, C., & Tamer, U. (2013). Rapid analysis of sugars in honey by processing Raman spectrum using chemometric methods and artificial neural networks. *Food Chemistry*, *136*(3–4), 1444–1452. doi:10.1016/j.foodchem.2012.09.064 PMID:23194547

Panche, A. N., Diwan, A. D., & Chandra, S. R. (2016). Flavonoids: An overview. *Journal of Nutritional Science*, *5*(e47), 1–15. doi:10.1017/jns.2016.41 PMID:28620474

Pangestika, N. W., Atmowidi, T., & Kahono, S. (2017). Pollen load and flower constancy of three species of stingless bees (Hymenoptera, Apidae, Meliponinae). *Tropical Life Sciences Research*, *28*(2), 179–187. doi:10.21315/tlsr2017.28.2.13 PMID:28890769

Pauly, A., Pedro, S. R. M., Rasmussen, C., & Roubik, D. W. (2013). Stingless bees (Hymenoptera: Apoidea: Meliponini) of French Guiana. In Pot-Honey (pp. 87–97). Springer.

Pedreschi, R., Hertog, M., Lilley, K. S., & Nicolai, B. (2010). Proteomics for the food industry: Opportunities and challenges. *Critical Reviews in Food Science and Nutrition*, *50*(7), 680–692. doi:10.1080/10408390903044214 PMID:20694929

Pedro, S. R. M., & Camargo, J. M. F. (2003). Meliponini neotropicais: O gênero Partamona Schwarz, 1939 (Hymenoptera, Apidae). *Revista Brasileira de Entomologia*, *47*, 1–117. doi:10.1590/S0085-56262003000500001

Ponnuchamy, R., Bonhomme, V., Prasad, S., Das, L., Patel, P., Gaucherel, C., Pragasam, A., & Anupama, K. (2014). Honey pollen: Using melissopalynology to understand foraging preferences of bees in tropical South India. *PLoS One*, *9*(7), 1–11. doi:10.1371/journal.pone.0101618 PMID:25004103

Pontis, J. A., da Costa, L. A. M. A., da Silva, S. J. R., & Flach, A. (2014). Color, phenolic and flavonoid content, and antioxidant activity of honey from Roraima, Brazil. *Food Science and Technology (Campinas)*, *34*(1), 69–73. doi:10.1590/S0101-20612014005000015

Prinfolosirea, M. E. C. M. O. (1965). Diferitelor Metode de Pregatire a Naterialului biologic. *Lucr Stiint Stat Cont Seri Apic*, *6*, 15–21.

Priyadarshan, P. M. (2011). *Biology of Hevea rubber*. Springer. doi:10.1079/9781845936662.0000

Pucciarelli, A. B., Schapovaloff, M. E., Kummritz, S., Señuk, I. A., Brumovsky, L. A., & Dallagnol, A. M. (2014). Microbiological and physicochemical analysis of yateí (Tetragonisca angustula) honey for assessing quality standards and commercialization. *Revista Argentina de Microbiologia*, *46*(4), 325–332. doi:10.1016/S0325-7541(14)70091-4 PMID:25576417

Puscas, A., Hosu, A., & Cimpoiu, C. (2013). Application of a newly developed and validated high-performance thin-layer chromatographic method to control honey adulteration. *Journal of Chromatography. A*, *1272*, 132–135. doi:10.1016/j.chroma.2012.11.064 PMID:23245847

Queiroz-Junior, C. M., Silveira, K. D., de Oliveira, C. R., Moura, A. P., Madeira, M. F. M., Soriani, F. M., Ferreira, A. J., Fukada, S. Y., Teixeira, M. M., Souza, D. G., & da Silva, T. A. (2015). Protective effects of the angiotensin type 1 receptor antagonist losartan in infection-induced and arthritis-associated alveolar bone loss. *Journal of Periodontal Research*, 50(6), 814–823. doi:10.1111/jre.12269 PMID:25753377

Ramón-Sierra, J. M., Ruiz-Ruiz, J. C., & de la Luz Ortiz-Vázquez, E. (2015). Electrophoresis characterization of protein as a method to establish the entomological origin of stingless bee honeys. *Food Chemistry*, 183, 43–48. doi:10.1016/j.foodchem.2015.03.015 PMID:25863608

Ranneh, Y., Ali, F., Zarei, M., Akim, A. M., Abd Hamid, H., & Khazaai, H. (2018). Malaysian stingless bee and Tualang honeys: A comparative characterization of total antioxidant capacity and phenolic profile using liquid chromatography-mass spectrometry. *Lebensmittel-Wissenschaft + Technologie*, 89, 1–9. doi:10.1016/j.lwt.2017.10.020

Rao, P. S., & Vijayakumar, K. R. (1992). Climatic requirements. In *Developments in crop science* (Vol. 23, pp. 200–219). Elsevier.

Rao, P. V., Krishnan, K. T., Salleh, N., & Gan, S. H. (2016). Biological and therapeutic effects of honey produced by honey bees and stingless bees: A comparative review. *Revista Brasileira de Farmacognosia*, 26(5), 657–664. doi:10.1016/j.bjp.2016.01.012

Rasmussen, C. (2008). Catalog of the Indo-Malayan/Australasian stingless bees (Hymenoptera: Apidae: Meliponini). *Zootaxa*, 1935(1), 1–80. doi:10.11646/zootaxa.1935.1.1

Ratnieks, F. L. W., Foster, K. R., & Wenseleers, T. (2006). Conflict resolution in insect societies. *Annual Review of Entomology*, 51(1), 581–608. doi:10.1146/annurev.ento.51.110104.151003 PMID:16332224

Revilla, G., Ramos, F. R., López-Nieto, M. J., Alvarez, E., & Martín, J. F. (1986). Glucose represses formation of delta-(L-alpha-aminoadipyl)-L-cysteinyl-D-valine and isopenicillin N synthase but not penicillin acyltransferase in Penicillium chrysogenum. *Journal of Bacteriology*, 168(2), 947–952. doi:10.1128/jb.168.2.947-952.1986 PMID:3096965

Ribeiro, M. de F., Wenseleers, T., Santos Filho, P. S., & Alves, D. A. (2006). Miniature queens in stingless bees: Basic facts and evolutionary hypotheses. *Apidologie*, 37(2), 191–206. doi:10.1051/apido:2006023

Rohlf, F. J. (1990). Morphometrics. *Annual Review of Ecology and Systematics*, 299–316.

Roselino, A. C., Santos, S. B., Hrncir, M., & Bego, L. R. (2009). Differences between the quality of strawberries (Fragaria x ananassa) pollinated by the stingless bees Scaptotrigona aff. depilis and Nannotrigona testaceicornis. *Genetics and Molecular Research*, 8(2), 539–545. doi:10.4238/vol8-2kerr005 PMID:19551642

Rosli, F. N., Hazemi, M. H. F., Akbar, M. A., Basir, S., Kassim, H., & Bunawan, H. (2020). Stingless bee honey: Evaluating its antibacterial activity and bacterial diversity. *Insects*, 11(8), 1–13. doi:10.3390/insects11080500 PMID:32759701

Roubik, D. W. (1983). Nest and colony characteristics of stingless bees from Panama (Hymenoptera: Apidae). *Journal of the Kansas Entomological Society*, 327–355.

Roubik, D. W. (1992). *Ecology and natural history of tropical bees*. Cambridge University Press.

Roubik, D. W. (2006). Stingless bee nesting biology. *Apidologie*, *37*(2), 124–143. doi:10.1051/apido:2006026

Roubik, D. W., Sakagami, S. F., & Kudo, I. (1985). A note on distribution and nesting of the Himalayan honey bee Apis laboriosa Smith (Hymenoptera: Apidae). *Journal of the Kansas Entomological Society*, 746–749.

Roulston, T. H., & Goodell, K. (2011). The role of resources and risks in regulating wild bee populations. *Annual Review of Entomology*, *56*(1), 293–312. doi:10.1146/annurev-ento-120709-144802 PMID:20822447

Sabir, A., Agus, A., Sahlan, M., & Agussalim. (2021). The minerals content of honey from stingless bee *Tetragonula laeviceps* from different regions in Indonesia. *Livestock Research for Rural Development, 33*(2).

Sagili, R. R., & Burgett, D. M. (2011). *Evaluating honey bee colonies for pollination: a guide for commercial growers and beekeepers*. Academic Press.

Sakagami, S F, Inoue, T., & Salmah, S. (1990). Stingless bees of central Sumatra. *Stingless Bees of Central Sumatra*, 125–137.

Sakagami, S. F. (1961). Bees of Xylocopinae and Apidae collected by the Osaka City University Biological Expedetion to Southeast Asia 1957-58, with some biological notes. *Nature and Life SE Asia, 1*, 409–444.

Sakagami, S. F. (1975). Stingless bees (excl. Tetragonula) from the continental Southeast Asia in the collection of Berince P. Bishop Museum, Honolulu (Hymenoptera, Apidae)(with 14 text-figures and 3 tables). 北海道大學理學部紀要, *20*(1), 49–76.

Sakagami, S. F. (1978). Tetragonula Stingless bees of the continental Asia and Sri Lanka (Hymenoptera, Apidae)(with 124 text-figures, 1 plate and 36 tables). 北海道大學理學部紀要, *21*(2), 165–247.

Sakagami, S. F., Inoue, T., & Salmah, S. (1990). Stingless bees of central Sumatra. *Stingless Bees of Central Sumatra.*, 125–137.

Sakagami, S. F., Yamane, S., & Inoue, T. (1983). Oviposition behavior of two Southeast Asian stingless bees, Trigona (Tetragonula) laeviceps and T.(T.) pagdeni. 昆蟲, *51*(3), 441–457.

Sakakura, T., Sakagami, Y., & Nishizuka, Y. (1982). Dual origin of mesenchymal tissues participating in mouse mammary gland embryogenesis. *Developmental Biology*, *91*(1), 202–207. doi:10.1016/0012-1606(82)90024-0 PMID:7095258

Salim, H. M. W., Dzulkiply, A. D., Harrison, R. D., Fletcher, C., Kassim, A. R., & Potts, M. D. (2012). Stingless bee (Hymenoptera: Apidae: Meliponini) diversity in dipterocarp forest reserves in Peninsular Malaysia. *The Raffles Bulletin of Zoology, 60*(1), 213–219.

Salmah, S. (1991). Incubation period in the Sumatra stingless bee Trigona (tetragonula) minangkabau. *Treubia, 30*, 195–201.

Salmah, S. (1989). *Tempat dan volume beberapa jenis lebah yang terdapat di Sumatera (Hymenoptera: Apidae)*. Seminar Dan Kongres Biologi Nasional IX Di Padang.

Samborska, K., & Czelejewska, M. (2014). The influence of thermal treatment and spray drying on the physicochemical properties of Polish honeys. *Journal of Food Processing and Preservation, 38*(1), 413–419. doi:10.1111/j.1745-4549.2012.00789.x

Sánchez-Moreno, C., Larrauri, J. A., & Saura-Calixto, F. (1998). A procedure to measure the antiradical efficiency of polyphenols. *Journal of the Science of Food and Agriculture, 76*(2), 270–276. doi:10.1002/(SICI)1097-0010(199802)76:2<270::AID-JSFA945>3.0.CO;2-9

Sanders, E. R. (2012). Aseptic Laboratory Techniques: Plating Methods (2022). *Journal of Visualized Experiments, 63*(63), e3064. doi:10.3791/3064 PMID:22617405

Saufi, N. F. M., & Thevan, K. (2015). Characterization of nest structure and foraging activity of stingless bee, Geniotrigona thoracica (Hymenopetra: Apidae; Meliponini). *Jurnal Teknologi, 77*(33), 69–74.

Saunders, M. E. (2018). Insect pollinators collect pollen from wind-pollinated plants: Implications for pollination ecology and sustainable agriculture. *Insect Conservation and Diversity, 11*(1), 13–31. doi:10.1111/icad.12243

Schwarz, H. F. (1948). Stingless bees (Meliponinae) of the western hemisphere. *Bulletin of the American Museum of Natural History, 90*, 1–546.

Seraglio, S. K. T., Silva, B., Bergamo, G., Brugnerotto, P., Gonzaga, L. V., Fett, R., & Costa, A. C. O. (2019). An overview of physicochemical characteristics and health-promoting properties of honeydew honey. *Food Research International, 119*, 44–66. doi:10.1016/j.foodres.2019.01.028 PMID:30884675

Shamsudin, S., Selamat, J., Sanny, M., Abd. Razak, S.-B., Jambari, N. N., Mian, Z., & Khatib, A. (2019). Influence of origins and bee species on physicochemical, antioxidant properties and botanical discrimination of stingless bee honey. *International Journal of Food Properties, 22*(1), 239–264. doi:10.1080/10942912.2019.1576730

Shehata, M. G., El Sohaimy, S. A., El-Sahn, M. A., & Youssef, M. M. (2016). Screening of isolated potential probiotic lactic acid bacteria for cholesterol lowering property and bile salt hydrolase activity. *Annals of Agricultural Science, 61*(1), 65–75. doi:10.1016/j.aoas.2016.03.001

Shimizu, K., & Morse, D. E. (2018). Silicatein: A unique silica-synthesizing catalytic triad hydrolase from marine sponge skeletons and its multiple applications. *Methods in Enzymology, 605*, 429–455. doi:10.1016/bs.mie.2018.02.025 PMID:29909834

Shipp, J. L., Whitfield, G. H., & Papadopoulos, A. P. (1994). Effectiveness of the bumble bee, Bombus impatiens Cr.(Hymenoptera: Apidae), as a pollinator of greenhouse sweet pepper. *Scientia Horticulturae*, *57*(1–2), 29–39. doi:10.1016/0304-4238(94)90032-9

Shivaramu, K., Sakthivel, T., & Reddy, P. V. (2012). Diversity and foraging dynamics of insect pollinators on rambutan (Nephelium lappacum L.). *Pest Management in Horticultural Ecosystems*, *18*(2), 158–160.

Šimúth, J., Bílíková, K., Kováčová, E., Kuzmová, Z., & Schroder, W. (2004). Immunochemical approach to detection of adulteration in honey: Physiologically active royal jelly protein stimulating TNF-α release is a regular component of honey. *Journal of Agricultural and Food Chemistry*, *52*(8), 2154–2158. doi:10.1021/jf034777y PMID:15080614

Singh, I., & Singh, S. (2018). Honey moisture reduction and its quality. *Journal of Food Science and Technology*, *55*(10), 3861–3871. doi:10.100713197-018-3341-5 PMID:30228384

Slaa, E. J., Chaves, L. A. S., Malagodi-Braga, K. S., & Hofstede, F. E. (2006). Stingless bees in applied pollination: Practice and perspectives. *Apidologie*, *37*(2), 293–315. doi:10.1051/apido:2006022

SNI. (2018). *Standar nasional Indonesia madu*. Badan Standarisasi Nasional.

Solayman, M., Islam, M. A., Paul, S., Ali, Y., Khalil, M. I., Alam, N., & Gan, S. H. (2016). Physicochemical properties, minerals, trace elements, and heavy metals in honey of different origins: A comprehensive review. *Comprehensive Reviews in Food Science and Food Safety*, *15*(1), 219–233. doi:10.1111/1541-4337.12182 PMID:33371579

Sommeijer, M. J., & De Bruijn, L. L. M. (1984). Social Behaviour of Stingless Bees:"Bee-Dances" by Workers of the Royal Court and the Rhythmicity of Brood Cell Provisioning and Oviposition Behaviour. *Behaviour*, *89*(3–4), 299–315. doi:10.1163/156853984X00434

Souza, B., Roubik, D., Barth, O., Heard, T., Enríquez, E., Carvalho, C., Villas-Bôas, J., Marchini, L., Locatelli, J., & Persano-Oddo, L. (2006). Composition of stingless bee honey: Setting quality standards. *Interciencia*, *31*(12), 867–875.

Suhaizan, L., Norezienda, A., Shamsul, B., Nurul, F. I., & Iffah, H. (2017). Composition and identification of pollen collected by stingless bee (Heterotrigona itama) in forested and coastal area of Terengganu, Malaysia. *Malaysian Applied Biology Journal*, *46*(3), 227–232.

Suhartatik, N., Cahyanto, M. N., Rahardjo, S., Miyashita, M., & Rahayu, E. S. (2014). Isolation and identification of lactic acid bacteria producing [Beta] glucosidase from Indonesian fermented foods. *International Food Research Journal*, *21*(3), 973–978.

Sujan, M. A., Atim, A. B., & Yaakob, A. M. (1984). Potensi penghasilan madu lebah di kawasan tanaman getah sebagai hasil samping. Siaran Pekebun.

Suntiparapop, K., Prapaipong, P., & Chantawannakul, P. (2012). Chemical and biological properties of honey from Thai stingless bee (Tetragonula laeviceps). *Journal of Apicultural Research*, *51*(1), 45–52. doi:10.3896/IBRA.1.51.1.06

Taha, E. K. A., & Alqarni, A. S. (2013). Morphometric and reproductive organs characters of Apis mellifera jemenitica drones in comparison to Apis mellifera carnica. *International Journal of Scientific and Engineering Research*, *4*(10), 411–415.

Tajuddin, I. (1986). Integration of animals in rubber plantations. *Agroforestry Systems*, *4*(1), 55–66. doi:10.1007/BF01834702

Taormina, P. J., Niemira, B. A., & Beuchat, L. R. (2001). Inhibitory activity of honey against foodborne pathogens as influenced by the presence of hydrogen peroxide and level of antioxidant power. *International Journal of Food Microbiology*, *69*(3), 217–225. doi:10.1016/S0168-1605(01)00505-0 PMID:11603859

Tarpy, D. R., Keller, J. J., Caren, J. R., & Delaney, D. A. (2011). Experimentally induced variation in the physical reproductive potential and mating success in honey bee queens. *Insectes Sociaux*, *58*(4), 569–574. doi:10.100700040-011-0180-z

Thakodee, T., Deowanish, S., & Duangmal, K. (2018). Melissopalynological analysis of stingless bee (Tetragonula pagdeni) honey in Eastern Thailand. *Journal of Asia-Pacific Entomology*, *21*(2), 620–630. doi:10.1016/j.aspen.2018.04.003

Thawai, C., Tanasupawat, S., Itoh, T., Suwanborirux, K., & Kudo, T. (2004). Micromonospora aurantionigra sp. nov., isolated from a peat swamp forest in Thailand. *Actinomycetologica*, *18*(1), 8–14. doi:10.3209aj.18_8

Tornuk, F., Karaman, S., Ozturk, I., Toker, O. S., Tastemur, B., Sagdic, O., Dogan, M., & Kayacier, A. (2013). Quality characterization of artisanal and retail Turkish blossom honeys: Determination of physicochemical, microbiological, bioactive properties and aroma profile. *Industrial Crops and Products*, *46*, 124–131. doi:10.1016/j.indcrop.2012.12.042

Tosi, E., Martinet, R., Ortega, M., Lucero, H., & Ré, E. (2008). Honey diastase activity modified by heating. *Food Chemistry*, *106*(3), 883–887. doi:10.1016/j.foodchem.2007.04.025

Trianto, M., & Purwanto, H. (2020). Morphological characteristics and morphometrics of stingless bees (Hymenoptera: Meliponini) in Yogyakarta, Indonesia. *Biodiversitas Journal of Biological Diversity*, *21*(6), 2619–2628. doi:10.13057/biodiv/d210633

Turkmen, N., Sari, F., Poyrazoglu, E. S., & Velioglu, Y. S. (2006). Effects of prolonged heating on antioxidant activity and colour of honey. *Food Chemistry*, *95*(4), 653–657. doi:10.1016/j.foodchem.2005.02.004

Vallianou, N. G., Gounari, P., Skourtis, A., Panagos, J., & Kazazis, C. (2014). Honey and its anti-inflammatory, anti-bacterial and anti-oxidant properties. *General Medicine (Los Angeles, Calif.)*, *2*(132), 1–5. doi:10.4172/2327-5146.1000132

Venturieri, G. C. (2004). *Criação de abelhas indígenas sem ferrão*. Embrapa Amazônia Oriental.

Venturieri, G. C. (2008). *Caixa para a criação de uruçu-amarela Melipona flavolineata Friese, 1900. In Embrapa Amazônia Oriental-Comunicado Técnico*. INFOTECA-E.

Veth, P. J. (1881). *Midden-Sumatra: reizen en onderzoekingen der Sumatra-expeditie uitgerust door het Aardrijkskundig genootschap, 1877-1879* (Vol. 1, Issue 1). Brill Archive.

Vit, P., Pedro, S. R. M., & Roubik, D. (2013). Pot-honey: A legacy of stingless bees. Springer Science & Business Media.

Vit, P., Bogdanov, S., & Kilchenmann, V. (1994). Composition of Venezuelan honeys from stingless bees (Apidae: Meliponinae) and Apis mellifera L. *Apidologie, 25*(3), 278–288. doi:10.1051/apido:19940302

Wanjai, C., Sringarm, K., Santasup, C., Pak-Uthai, S., & Chantawannakul, P. (2012). Physicochemical and microbiological properties of longan, bitter bush, sunflower and litchi honeys produced by Apis mellifera in Northern Thailand. *Journal of Apicultural Research, 51*(1), 36–44. doi:10.3896/IBRA.1.51.1.05

White, J. W., & Doner, L. W. (1980). Honey composition and properties. *Beekeeping in the United States Agriculture Handbook, 335*, 82–91.

Wikler, M. A. (2006). Methods for dilution antimicrobial susceptibility tests for bacteria that grow aerobically: Approved standard. *Clsi (Nccls), 26*, M7–A7.

Wilson, E. O. (1971). *The insect societies.* Cambridge, Massachusetts, USA, Harvard University Press [Distributed by…. Wittmann, D. (1985). Aerial defense of the nest by workers of the stingless bee Trigona (Tetragonisca) angustula (Latreille)(Hymenoptera: Apidae). *Behavioral Ecology and Sociobiology, 16*(2), 111–114.

Won, S.-R., Lee, D.-C., Ko, S. H., Kim, J.-W., & Rhee, H.-I. (2008). Honey major protein characterization and its application to adulteration detection. *Food Research International, 41*(10), 952–956. doi:10.1016/j.foodres.2008.07.014

Woyke, J. (1971). Correlations between the age at which honeybee brood was grafted, characteristics of the resultant queens, and results of insemination. *Journal of Apicultural Research, 10*(1), 45–55. doi:10.1080/00218839.1971.11099669

Yadav, S., Stow, A. J., Harris, R. M. B., & Dudaniec, R. Y. (2018). Morphological variation tracks environmental gradients in an agricultural pest, Phaulacridium vittatum (Orthoptera: Acrididae). *Journal of Insect Science, 18*(6), 1–13. doi:10.1093/jisesa/iey121 PMID:30508202

Yang, S.-C., Lin, C.-H., Sung, C. T., & Fang, J.-Y. (2014). Antibacterial activities of bacteriocins: Application in foods and pharmaceuticals. *Frontiers in Microbiology, 5*, 1–14.

Yap, S. K., Chin, N. L., Yusof, Y. A., & Chong, K. Y. (2019). Quality characteristics of dehydrated raw Kelulut honey. *International Journal of Food Properties, 22*(1), 556–571. doi:10.1080/10942912.2019.1590398

Yoshikawa, K. (1969). Preliminary report on entomology of the Osaka City University 5th Scientific Expedition to Southeast Asia 1966-With descriptions of two new genera of stenogastrine wasps by J. van der Vecht. *Nature and Life in Southeast Asia, 6*, 153–182.

Yusop, S. A. T. W., Asaruddin, M. R., Sukairi, A. H., & Sabri, W. M. A. W. (2018). Cytotoxicity and Antimicrobial Activity of Propolis from Trigona itama Stingless Bees against Staphylococcus aureus and Escherichia coli. *Indonesian Journal of Pharmaceutical Science and Technology*, *1*(1), 13–20.

Yusop, S. M., O'Sullivan, M. G., Kerry, J. F., & Kerry, J. P. (2010). Effect of marinating time and low pH on marinade performance and sensory acceptability of poultry meat. *Meat Science*, *85*(4), 657–663. doi:10.1016/j.meatsci.2010.03.020 PMID:20416811

Zamora, G., Arias, M. L., Aguilar, I., & Umaña, E. (2013). Costa Rican pot-honey: its medicinal use and antibacterial effect. In *Pot-Honey* (pp. 507–512). Springer. doi:10.1007/978-1-4614-4960-7_37

Zeedan, E. W. M. (2002). *Studies on certain factors affecting production and quality of queen honeybees (Apis mellifera L.) in Giza region* [Sc. Thesis]. Fac. Agric. Cairo Univ.

Zerdani, I., Abouda, Z., Kalalou, I., Faid, M., & Ahami, M. (2011). The Antibacterial Activity of Moroccan Bee Bread and Bee-Pollen (Fresh and Dried) against Pathogenic Bacteria. Res. *Journal of Microbiology (Seoul, Korea)*, *6*, 376–384.

Zhang, Y., Song, Y., Zhou, T., Liao, X., Hu, X., & Li, Q. (2012). Kinetics of 5-hydroxymethylfurfural formation in chinese acacia honey during heat treatment. *Food Science and Biotechnology*, *21*(6), 1627–1632. doi:10.100710068-012-0216-9 PMID:31807335

Zulkhairi Amin, F. A., Sabri, S., Mohammad, S. M., Ismail, M., Chan, K. W., Ismail, N., Norhaizan, M. E., & Zawawi, N. (2018). Therapeutic properties of stingless bee honey in comparison with european bee honey. *Advances in Pharmacological Sciences*, *2018*, 2018. doi:10.1155/2018/6179596 PMID:30687402

About the Contributors

Jumadil Saputra is a Ph.D. holder and works as a senior lecturer in the Department of Economics, Faculty of Business, Economics, and Social Development, Universiti Malaysia Terengganu, Malaysia. Currently, he has supervised and graduated 21 undergraduate students for completing the final year project and a total of 8 undergraduate students in progress completing their final year project. For the postgraduate student, he has supervised 5 master students, and 45 Ph.D. students with 9 students are role as the main supervisor. Further, there are 20 students, supervisory, and 15 internship students. He is a co-researcher for 12 grants, which is the university, national and international levels. Besides, starting from 2018 to 2020, he has published 265 articles journal and proceedings Scopus and 27 WoS articles journal and proceedings indexed. His research areas are Quantitative Economics (Microeconomics, Macroeconomics, and Economic Development), Econometrics (Theory, Analysis, and Applied), Islamic Banking and Finance, Risk and Insurance, Takaful, i.e., financial economics (Islamic), mathematics and modeling of finance (Actuarial), Qualitative and Quantitative Research.

* * *

Ali Agus is a lecturer at the Faculty of Animal Science, Universitas Gadjah Mada, Indonesia.

Wahizatul Azmi's research interests are mainly on the diversity and ecology of insects, pest and disease management and environmental biology. Her area of specialization is analyzing and quantifying of biodiversity and community structure of insects, evaluating aquatic insects as bio-indicator of water quality and insects-plant interactions, especially on Coleopteran pests and stingless bees. Her current main research activities are mainly on developing a biocontrol agent using entomopathogenic fungus, Metarhizium aniopliae which has been nano-formulated as a control strategy against an invasive coconut pest, Rhynchophorus ferrugineus (Coleoptera: Dryophthoridae). Besides that, she is also actively investigating on in-vitro rearing of queen Indo-Malaya stingless bee, Heterotrigona itama (Hymenoptera: Apidae) as well as the pollination ecology of the native stingless bees.

Nor Hazwani binti Mohd Hasali has a master's degree in science (Biotechnology) from International Islamic University Malaysia (2014) and a Ph.D in biotechnology (Food Microbiology) from University Malaysia Terengganu (2020). Her research involves food processing, food technology, and animal studies. She also has experience with analytical instruments for natural products.

Norizah Sarbon is a PhD student at Universiti Malaysia Terengganu.

Index

A

Antimicrobial Activity 29, 53, 56, 61, 63, 65, 137, 141, 144-147, 149-151, 188, 222

Antioxidant Activity 2, 23, 34, 69, 76-77, 79, 136, 138-139, 175, 177, 179-180, 182-183, 185-186, 188

Apini Tribe 104-105, 112, 207

Authenticity 5, 35, 78, 105-107, 116, 122, 149

B

Bao Technology 81-82

Bicol Region 81, 101

Bile Salts 142, 147-148

Biomarker 104-105

Breeding 28, 142, 190

C

Characters 153, 156, 204

Chili 41-45, 50, 72

Cucumbers 41-42, 48-49, 51

E

Ecosystem 41, 43, 50, 154, 169

G

Geniotrigona thoracica 107, 121-123, 137, 139, 175, 186, 218

H

Haemolytic Activity 147

Hetero-Cyclic Amine (HCA) 1-3, 10-12, 17, 22-23

Heterotrigona Itama 36, 39, 41-45, 50, 67, 70, 105, 107, 121-123, 137, 141, 159, 172, 174-175, 177, 186, 189, 207, 214, 218

Honey 1, 3-7, 11-12, 19-20, 22-29, 31-40, 51, 54-80, 82, 89-90, 93-94, 96-98, 100-102, 104-109, 111-152, 158-160, 166, 170-171, 173, 176, 179-181, 183, 188, 191-193, 199, 201, 203-204, 206-209, 211-212, 214-217, 219-221

Honey Bees 35, 51, 66, 119, 135, 160, 170-171, 176, 183, 188, 203, 206-208, 211-212, 214-216

L

Lepidotrigona 53, 70, 122, 175, 214

M

Marinade 1-3, 7-9, 11, 14, 16, 19-23, 27-28, 30, 33-34, 36, 38, 40, 52

Meliponiculture 54, 65, 70-73, 75-77, 82, 137, 150, 159, 169, 172, 175, 177, 179, 186, 190-192, 201, 203, 207

Meliponini 41, 51, 65-67, 70, 78-80, 101-102, 104-105, 107, 112, 116, 136, 139, 151, 153, 157, 200, 202-203, 207, 220-221

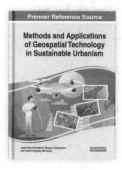
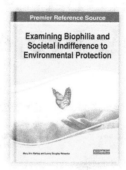

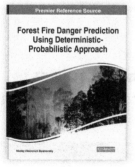

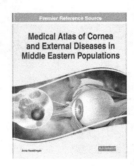

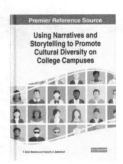

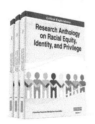
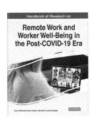

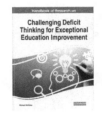

CPSIA information can be obtained
at www.ICGtesting.com
Printed in the USA
BVHW021409110123
656080BV00004B/148